GW00374386

THE TV VET SHEEP BOOK

£3.70

To Alex Wilson, MRCVS,
Veterinary Investigation Officer in charge of
the superb veterinary laboratory at the
West of Scotland Agricultural College, Auchincruive,
with sincere appreciation for his spontaneous
and generous help.

THE TV VET SHEEP BOOK

Recognition and Treatment of Common Sheep Ailments

By
THE TV VET

FARMING PRESS LIMITED
FENTON HOUSE WHARFEDALE ROAD IPSWICH SUFFOLK

FIRST PUBLISHED 1972
SECOND EDITION (Revised) 1973
SECOND IMPRESSION 1975

ISBN 0 85236 039 8

This book is set in 'Monophoto' Times 10pt on 11pt and is printed in Great Britain by Cox & Wyman Ltd, London, Fakenham and Reading

Contents

Acknowledgement is made to Dr J. P. Crowley, MVB, PhD, BAgSc, MRCVS, Head of Dept. of Animal Physiology, The Agricultural Institute, Dunsinea, Castleknock, Co. Dublin, Eire, for the photograph reproduced on page 31, and to Mr W. A. Watson, F, BVSc, for the photograph of Bright Blindness on page 149.

Foreword

By Dr J. A. WATT, MRCVS, PhD, BSc, Head of
Veterinary Department and Director of Veterinary Research
Edinburgh and East of Scotland School of Agriculture

AS ONE who has long been involved in diagnosing disease and advising farmers and shepherds on health and flock management, it gives me considerable pleasure to write a Foreword to this welcome addition to the subject of health in the sheep flock. The approach is one which has been followed by the author in his previous work, that is, the liberal use of photographs to illustrate recognisable conditions and operations, accompanied by a clear, condensed text which demonstrates the author's practical experience.

The format of illustration and readable text should appeal in particular to stock-owners and students, while avoiding the pitfall of producing 'do it yourself' experts. The importance of the sheep in agriculture tends to be underestimated, but in some areas in Britain they are of major and increasing importance in the livestock industry.

There can be few species in which so much attention has been directed towards the conception of flock protection and health, as opposed to conditions in the individual animal, and this wealth of epidemiological information has of itself generated the problem of dissemination and application of this knowledge. It has long been recognised that this educational field, *i.e.* the application of results of research on the farm, is a difficult one, and for this reason material presented as a readable and practical reference work performs a valuable function.

This book, then, fills an important gap in the series so ably presented by the author, and its unique form of presentation will result in the dissemination of information in many quarters where the stereotyped text-book has little appeal.

South Queensferry,
West Lothian.

J. A. WATT

Author's Preface

ONE BOTTLE of lambing oils, one bottle of fever drink, a piece of rolled-up fencing wire, pliers and hammer. This was my equipment to fight disease and losses in a Scottish hillside flock during the early spring of 1938—this and the limited experience of two previous lambing seasons. The shepherd was an alcoholic, and I had to cope almost entirely on my own. The wire, pliers and hammer were my 'stitching tackle'.

Looking back with present-day knowledge, it seems miraculous that my losses were not catastrophic: in fact they were comparatively slight. The only possible explanation must have been the natural resistance of a completely self-contained flock.

Today things are quite different. The modern flockmaster can embark on a lambing season with an armoury of preventive drugs, so comprehensive that it should be a disgrace to lose a single ewe or lamb. With his local knowledge and his hypodermic syringe he should be as inviolate as any Dr Kildare.

It is my hope that this work will serve as a simple, easily understood reference book not only for such flockmasters, but also for generations of agricultural and veterinary students and for veterinary surgeons in practice.

I acknowledge with sincere thanks the work and publications of the veterinary scientists, without reference to which it would not have been possible to present such a comprehensive disease picture. And once again, I express my appreciation and thanks to my photographer, Tony Boydon, and his colleague, Richard Perry, who drew the diagrams.

METABOLIC DISEASES

Metabolic Diseases

PERHAPS THE term 'disorders' is more appropriate since there are no germs involved in metabolic diseases.

They are produced by disturbances in the animal's metabolism. *i.e.* the conversion of food into the appropriate substances to keep the normal body functions going. The main organ concerned in metabolism is the liver which could be described as the 'factory of the body'. There the simple products of intestinal digestion are elaborated and built up before being transported to the various parts of the anatomy to provide heat and energy, and to build up and replace muscle, brain, etc. In order to perform this function the liver must be constantly supplied not only with the basic products of digestion—fatty and amino acids—but also with minerals, trace elements and vitamins.

Also the 'factory' occasionally has to play a part in breaking down stored food chiefly body fat.

The metabolic diseases that concern us most in sheep are, first, pregnancy toxaemia, caused by excess ketones forming in the blood during the break down of body fat, secondly, hypomagnesaemia, due to a shortage of the mineral magnesium and, thirdly, lambing sickness which, like milk fever in the cow, is caused by a deficiency of calcium.

Strictly speaking, however, CCN should be included under this heading, as also should swayback, pine or cobalt deficiency, hypovitaminosis E, and phosphorus deficiency.

1

Pregnancy Toxaemia

IN THE simplest possible language pregnancy toxaemia is brought about by inadequate feeding, or to put it even more bluntly, by starvation. This is why the disease occurs when there is a deterioration in the quality and quantity of the diet during the last 2 months of pregnancy.

For example, it is seen in flocks maintained on pasture *(photo 1)* with little or no supplementary feed, even though occasionally on a good pasture during an open winter only a few cases may occur. Or, again, it is seen in severe weather where frosts and snows drastically reduce or completely cover up the available herbage.

How It Occurs

In late pregnancy, especially if twin lambs are present, the ewe requires a good food intake to keep pace with the growth of the lambs—naturally. If she is not getting enough food by the mouth, she has to turn to her own food reserves which are stored in the form of sugar or glycogen in the liver and muscles, and in the form of fat throughout the body.

The hungry ewe quickly uses up all the sugar reserves and this causes a drop in the blood sugar content. She then turns to the body fat; this is broken down in the liver, and during the breaking-down process, poisonous substances called ketones are formed *(see diagram, page 16)*. These accumulate in the bloodstream and produce an effect not unlike that caused by excessive alcohol in humans.

Symptoms

The first sign is the ewe's disinclination to move, quite distinct from any disability. The animal is often described as being 'stupid' and this symptom of stupidity,

1

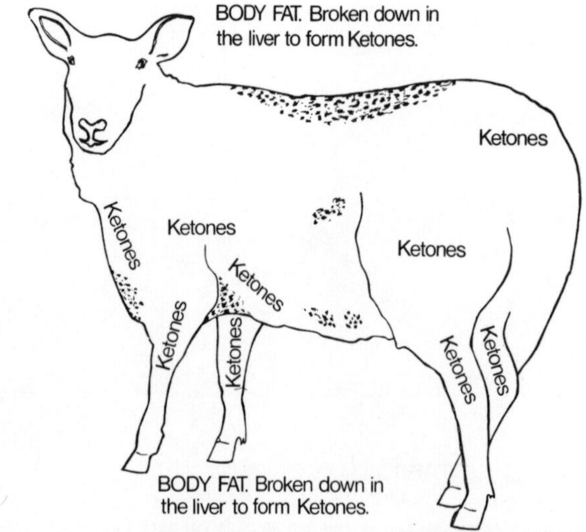

These accumulate in the bloodstream to cause PREGNANCY TOXAEMIA

together with the history of pregnancy, is virtually diagnostic.

In this state the ewe will often face, and even try to fight off, the sheepdog *(photo 2)*.

Later the ewe's head is carried in an unnatural position, usually to one side *(photo 3)*, and it may be held high or dropped low.

The gait becomes staggering and uncertain and later still there is a disturbance of vision and even blindness *(photo 4)*. This is why in many parts of the country the condition is described as 'snow blindness'.

The temperature is usually normal, but the appetite is completely lost, and the patient is invariably constipated.

In about 1 to 2 days the ewe is unable to stand; she becomes progressively comatose and dies in from 1 to 6 days.

2

3

Treatment

Treatment of cases is very unsatisfactory; unless the ewe is close to lambing the mortality rate is about 90 per cent.

My own rule of treatment is: If she is approximately within 1 week of lambing, I treat her by giving cortisone and vitamin B injections *(photo 5)* together with substances like treacle and glycerine by the mouth. A spectacular recovery very often follows the birth of

4

5

the lambs, though often the lambs are born dead.

If, however, lambing is some way off (usually the best way to estimate this is by examining the udder *(photo 6)*, then I recommend loading up the ewe immediately and taking her to the butcher.

Quite obviously, it is much better not to have to treat pregnancy toxaemia, and in the light of present-day knowledge, any intelligent shepherd can prevent it.

Prevention

The simple clue to the virtual elimination of pregnancy toxaemia in sheep lies in the feeding, particularly during the last 8 weeks of pregnancy when the major part of the foetal growth takes place.

Demand for extra food during this period depends on the number of lambs the ewe happens to be carrying. Since there is no way in which this can be assessed with certainty (except of course by expensive X-rays), the safest plan is to feed on the presumption that all the ewes are carrying twins.

6

7

The most important feed of all is good leafy hay, but this must always be provided *ad lib (photo 7)*. Again and again farmers moan that their ewes don't eat the hay they put out. This is nonsense. They don't eat non-stop like cattle or horses, but they nibble away at regular intervals and over the 24 hours pack in an adequate supply, provided (and this leads me to my second, most important, point) ample fresh water is available at all times.

The pregnant ewe needs plenty of water, as does the lactating ewe and, for that matter, all other sheep. The stupid idea that sheep don't drink water is still being bandied about, often by apparently intelligent countrymen.

Good hay and plenty of water, *ad lib*, could be all that is required if the ewes are housed. In fact, in housed sheep, top quality hay will provide sufficient for the full growth of a single lamb and most average-sized twins. But if you rely on hay alone, make sure you have it tested for quality—only the best is good enough.

If the ewes are outside—as most of them are—then it is a different matter. Much of the hay ration is used up in keeping the ewe's body heat normal and a supplementary diet of concentrates becomes vital *(photo 8)*.

8

The time to start the concentrates is 8 weeks before lambing, when the foetus or foeti really start to grow.

I've found over the years that the quality of concentrates depends largely on the type of land and pasture. If the ewes have been on good pasture, then the concentrate ration should be at least as rich in protein, probably 1 or 2 per cent richer than the best grass on the farm.

If the land and pasture are comparatively poor, then a cereal concentrate of crushed oats and flaked maize will do admirably, *but keep barley out of the ration*. It has been my experience that barley can do as much, if not more, harm to pregnant ewes as it does to cattle.

As an additive I always recommend diluted treacle. It makes the feed attractive and succulent and has the obvious benefit of being ketogenic providing readily assimilable carbohydrate.

In all hypomagnesaemia areas, added magnesium is vital. I advise feeding at the rate of $\frac{1}{4}$ oz of calcined magnesite a head each day (sheep find it most edible if wet beet pulp can be included in the ration), and I feed it right on to the middle of June.

As for other minerals, their necessity depends on the land and on the standard of farming. Certainly in my comparatively rich, well-farmed area the only trace element I advise is the odd 5 per cent copper lick on farms with a history of swayback. There is no doubt that copper in any quantity is highly dangerous to all sheep, and the weak lick is the only safe way I've found of feeding it.

It is my opinion that considerable amounts of money are dissipated in feeding minerals where they are not necessary. After all, the good hay, which I stipulated at the start, contains all the minerals and trace elements in the correct quantities and in an easily assimilated form. Excess minerals are not only wasteful, but they can also damage the liver and intestinal tract of the ewe and lead to diarrhoea and deaths.

9

Lastly, the quantity. Having compiled the ration carefully in consultation with your veterinary surgeon or your local advisory officer, and having added magnesium where necessary and any other vital element which is known to be missing from the soil and pasture of your respective farm, start 8 weeks before lambing with $\frac{1}{2}$ lb of your concentrate mixture for each ewe every day *(photo 9)* (in addition to the *ad lib* hay and water) and gradually increase the quantities, *i.e.* keep the ewes on an ascending plane of nutrition.

At 4 weeks before lambing feed 1–$1\frac{1}{2}$ lb per head a day, and at full term feed $2\frac{1}{2}$–3 lb according to the breed and size of your ewe flock. In a particularly severe winter it will pay to add 5 or 10 per cent of animal protein (fish meal or meat-and-bone meal) to the ration during the last 3 or 4 weeks.

Do all this and pregnancy toxaemia will occur only if or when you allow the water supply to freeze. *So never forget in severe weather always break the ice several times daily (photo 10).*

10

2
Hypomagnesaemia

HYPOMAGNESAEMIA IS common nearly everywhere, especially in lowland sheep *(photo 1)*. Nowadays it is probably the greatest potential danger to all flocks.

The syndrome may appear under several circumstances. It can occur after a journey from high ground to lowland pastures, due probably to a 'transit tetany', with the magnesium deficiency appearing as a direct result of the travelling.

Hypomagnesaemia can also break out in a flock during a sudden cold spell, no doubt because of the extra strain placed on the sheep's metabolic rate.

Most cases, however, arise in exactly the same way as in cattle—*i.e.* by a direct shortage of the mineral magnesium. Like cattle, the sheep magnesium reserves are stored on the crystalline surfaces of the framework of the bones, especially the ribs and vertebrae.

In young sheep the bony framework is open; the magnesium reserves are readily available and may last for 40 to 50 days—much longer than they do in older animals, where the bone structure is more compact. In fact, in adult sheep the magnesium reserves may keep them going for no more than 4 or 5 days. If the flock is on a low plane of nutrition, the magnesium reserves are soon exhausted.

On a good diet, however, hypomagnesaemia can, and does, occur quite commonly. Then it is probably triggered off by the hay *(photo 2)* (or silage, if fed) having been made early from young artificially flushed grasses.

In any case, even if symptoms do not appear during the winter, there is always a gradual natural decline in the magnesium reserves, and consequently all the sheep are lowest in magnesium in the early spring.

If they are then turned out on to a young, rapidly growing pasture which owes its early growth to artificials, the sheep can start to die by the score.

It appears that the artificials, especially potash, lock the magnesium in the soil

1

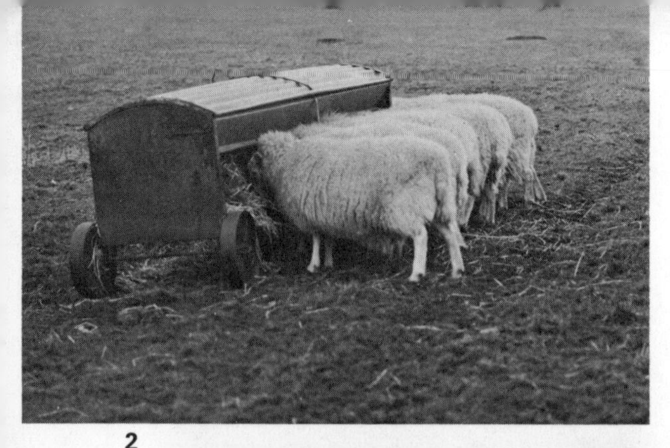

2

3

and hinder its uptake, or that the high nitrogen content of the artificially boosted grasses inhibits absorption in the ewe's intestinal tract.

Symptoms

In my experience—usually sudden death. Spotted in the early stages the sheep may stagger about; it shows a hyperexcitability with twitching muscles and grinding teeth and, if untreated, soon falls down in a fit *(photo 3)*—kicking and frothing at the mouth before becoming comatose and dying.

Treatment

If caught reasonably early—even while the animal is in a fit—treatment can be spectacularly successful. Often I have pumped 100 cc of magnesium solution *(photo 4)* into a prostrate kicking ewe and within an hour have been unable to spot any difference between her and the rest of the flock. But, for such results, treatment must be given early because

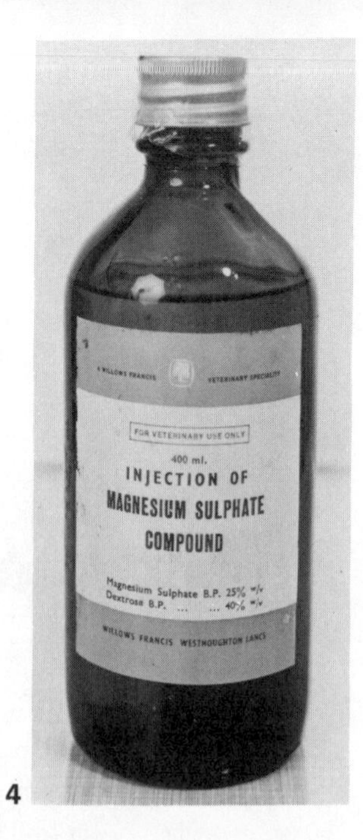

4

after a comparatively short time, haemorrhages occur in the brain and heart. When that happens, magnesium injections are useless.

Personally I like to use a 20-cc syringe and inject the ewe under the skin in five different places *(photo 5)*, coating the hypodermic needle with antibiotic between each one and massaging the sites to disperse the solution. In this way a much more rapid absorption and effect are obtained.

If the ewe is dead or dies, always get

5

your veterinary surgeon to perform a post-mortem examination. The scientists say that hypomagnesaemia post-mortems reveal little, but I have found over the years that one common feature, which I have come to regard as diagnostic, has been well-marked haemorrhages in the substance of the heart muscle.

Prevention

If the flock is at all suspect, it might be a good idea to have blood samples taken from say 10 per cent and checked for magnesium levels. However, I prefer to crash on with preventative therapy at the first sign of loss, and I think most practising veterinary surgeons do likewise. In fact I'm sure that the majority, like myself, will have the hypomagnesaemia farms ear-marked and will advise prevention at all times.

Luckily most hypomagnesaemia cases are found during the winter and early spring. This is fortunate because at these times we can control it by feeding the magnesium. This can be done in several ways: (i) as a magnesium-rich mineral, which must contain at least 60 per cent magnesium to be effective, (ii) as sheep cubes or nuts containing magnesium, or (iii) as magnesium acetate in molasses, dispensed in freely available ball licks.

I prefer to feed $\frac{1}{4}$ oz of calcined magnesite *(photo 6)* per head per day to the entire flock. The magnesite should be well mixed up with the concentrates, and *ad lib* hay and water should be provided also. Where possible, wet beet pulp or treacle added to the ration will make the calcined magnesite more palatable.

As a long-term policy, especially on arable and certain hill farms but in all the hypomagnesaemia areas, I advise the routine use of magnesian limestone instead of ordinary lime as a pasture

6

dressing, but make sure that it contains at least 10 per cent of magnesium. Spread at $2\frac{1}{2}$ tons per acre, magnesian limestone will increase the magnesium content of the pastures by as much as 25 per cent. The ideal dressings of course would be calcined magnesite (10 cwt to the acre) or Keiserite (100 lb to the acre), but both these are expensive.

One other commonsense prevention, especially in the spring, is to change the flock on to another pasture, preferably an old-established turf which has not had any quantity of artificials applied *(photo 7)*.

There is no doubt that when the disease flares up in a young ley, deaths will often cease dramatically if you do nothing else but simply shift the flock on to a permanent pasture. It seems that when magnesium is available, sheep can utilise it quickly.

If a client insists on maintaining an intensive grass programme, I advise alternating the grazing of the young leys with the grazing of older permanent pasture, at least until the end of June. These rest periods allow the grasses in the leys to take up a percentage of the magnesium.

No doubt the eventual answer to hypomagnesaemia will be the discovery of the precise trigger factor which inhibits the magnesium uptake.

7

3
Lambing Sickness

WHEREAS HYPOMAGNESAEMIA can be seen at any time, lambing sickness usually occurs either just before, during, or after lambing *(photo 1)*. The condition is identical to milk fever in the cow. Older ewes are more commonly affected.

Cause

It is caused by calcium deficiency. Considerable quantities of calcium are required to make the bones and teeth of the lambs. During and after the birth of the lambs the udder fills up with milk, which contains a fair percentage of calcium. Occasionally the ewe's thyroid gland cannot cope with this sudden additional demand and lambing sickness is the result.

Predisposing Causes

Fatigue—for example, following the gathering in of hill ewes. Sudden changes in feeding can trigger it off as can also starvation.

Symptoms

The ewe may drop dead or fall flat out when being driven, but usually she goes off her food and her ears become ice-cold. The muscles start to tremor, the hind legs stiffen and she staggers about like a drunken man. She then goes down and is unable to rise. The breathing is accelerated and, if untreated, bloat sets in. The temperature is normal and constipation is the rule.

Treatment

A spectacular response is obtained with intravenous or subcutaneous injections of a 20 per cent solution of calcium borogluconate. Again, 100 cc is the dose and when given subcutaneously is best injected at five different sites using a 20-cc syringe *(photo 2)*, coating the hypodermic needle and the sites with antibiotic.

Prevention

Unfortunately there is no panacea for the

1

2

prevention of lambing sickness. Where cases are prevalent it is wise to check the lime content of the pasture. Recently, after a severe run of trouble on my own farm, I found one of my sheep pastures required 4 tons to the acre. On hill sheep farms, however, this may not be practicable.

There are certain other commonsense procedures, *e.g.* avoid sudden changes of feeding in late pregnancy. And do not drive the ewes too much immediately prior to or immediately after lambing.

All shepherds, especially those on remote hill farms, should be provided with the necessary calcium solution, 20-cc syringe and needles *(photo 3)* so that they can treat suspect cases immediately.

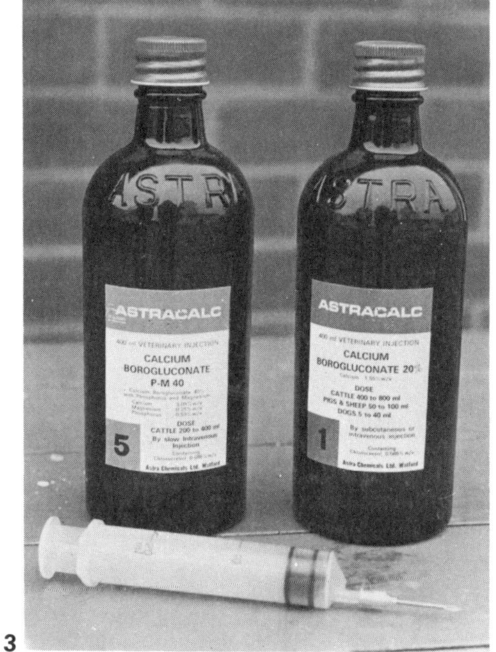

3

4
Cerebrocortical Necrosis

THIS DISEASE, known as CCN, is a nervous disease of ewes and lambs, associated with necrosis or death of part of the brain *(photo 1)*.

Causes

The precise cause is not known, but research work seems to indicate that it is due to a thiamine (vitamin B_1) deficiency caused by some metabolic disturbance. Certainly many cases, if spotted early enough, respond to treatment with injections of vitamin B_1.

Symptoms

In the early stages the affected ewes and lambs wander round aimlessly; they may walk in continuous circles or may just stand motionless. They also appear to be blind.

After a few hours they pitch forward on their side or brisket, throw their heads back *(photo 2)* and kick their legs as though in a fit. The legs often stiffen, but any sudden noise will trigger off the violent leg kicking.

There are two other conditions which can produce almost identical symptoms, viz. lead poisoning and hypomagnesaemia, though I have found lead poisoning to be extremely uncommon in sheep. Nonetheless a specific diagnosis can only be made on post-mortem examination in a laboratory.

1

2

Treatment

All suspect cases should be injected with vitamin B_1 immediately, either intravenously or intramuscularly *(photo 3)*; at the same time it is wise to provide a cover of long-acting antibiotic to prevent meningitis or other complications.

Since prompt differential diagnosis in the field is virtually impossible, it is wise to inject magnesium and to give a solution of magnesium sulphate by the mouth—both will take care of any magnesium deficiency, and the oral magnesium sulphate will act as the antidote to possible lead poisoning.

Prevention

Once the condition has been established on a farm, a vitamin supplement containing thiamine should be fed to the entire flock.

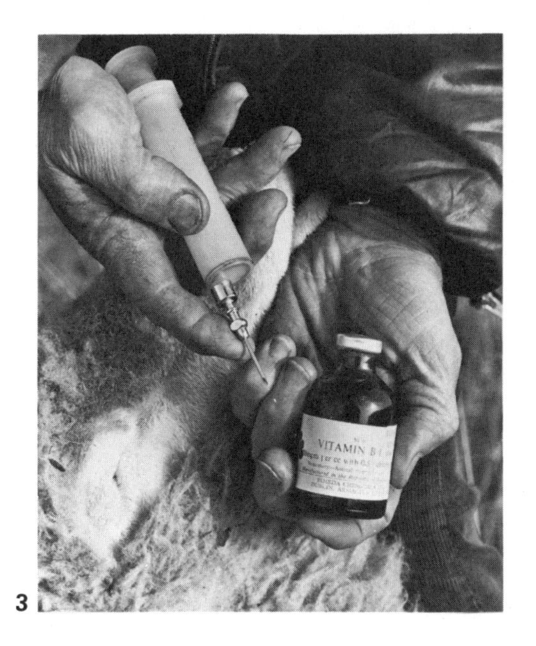

3

5
Swayback

SWAYBACK IS a disease of new-born lambs *(photo 1)*, though in mild cases the typical symptoms may not be readily apparent for a number of days or even weeks.

Cause

It is caused by a copper deficiency in the ewe's diet, producing a low blood copper level. This low level often occurs where there is no copper deficiency in the pastures. It would appear, therefore, that there is an undiscovered factor which inhibits the uptake of the copper. One predisposing cause appears to be over-liming of the pastures.

Symptoms

When the blood copper of the ewe is low, the lambs inside the mother suffer a varying amount of damage to the brain cells *(photo 2)*.

The brain damage usually manifests itself at birth by the lamb having difficulty in standing or in controlling its hind-quarters. However, in milder cases, the symptoms may not appear until the lamb is a few weeks old; then the affected lamb may only sway almost imperceptibly (hence the name swayback) *(photo 3)*. In fact, in very mild cases the hind-quarter weakness may only be seen when the lamb is chased.

1

2

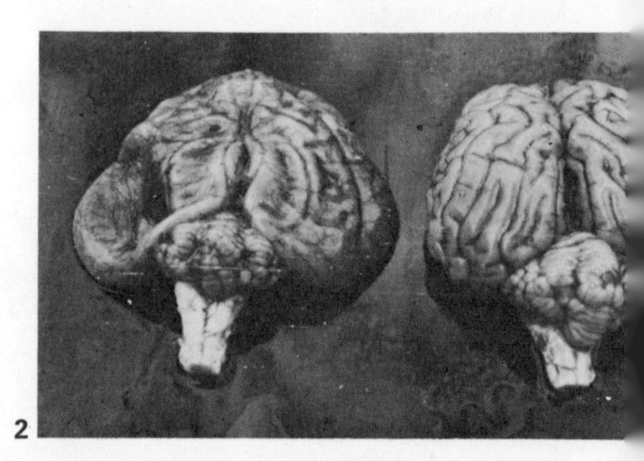

3

4

5

Treatment

There is no satisfactory treatment. Attempts at injecting or dosing with copper are dangerous because all sheep, and especially lambs, are highly susceptible to copper and may rapidly develop signs of copper poisoning; and copper poisoning is more severe and more rapidly fatal than swayback.

Chance Of Recovery?

It has been my experience that where the symptoms are severe, the lambs just don't get better. In fact, the swaying becomes progressively worse as the weight of the lamb increasès *(photo 4)*. However, I have seen some attacks mild enough to allow the lamb to be fed out in the normal way. My advice, therefore, is: where the lamb can walk it is wise to give it a chance.

Prevention

At all times avoid the over-liming of sheep pastures.

Provide 5 per cent copper licks to the ewe flock before and during pregnancy *(photo 5)*. The licks appear safe and are highly effective.

Copper may be injected as a preventative but only under strict veterinary supervision.

6
Pine

PINE IS a chronic wasting disease of sheep. In some respects it is similar to pernicious anaemia in humans.

Cause
The basic cause is a shortage or more usually a complete lack of the trace element cobalt. The trigger factor, however, is the inability of the sheep to manufacture vitamin B_{12} in the rumen because of the complete absence of cobalt.

Symptoms
Growing lambs fail to thrive in summer despite regular dosing against internal parasites. The older sheep have dull, dry fleeces *(photo 1)* and become stunted and handle badly. If the condition is not diagnosed, emaciation continues until the sheep can't stand, goes into a coma and dies.

Treatment
Fortunately the pine areas of Great Britain are now well known and prevention is more the rule. However, affected sheep respond rapidly to vitamin B_{12} injections plus dosing with a cobalt bullet *(photo 2)*. The cobalt bullet stays in the sheep's first stomach for about 6 months and keeps supplying the small amounts of cobalt necessary.

1

2

Prevention
Where pine is suspected, it is always wise to have the diagnosis confirmed by a research institute.

Prevention is not always easy, especially on hill land. Cobalt can be spread on the pastures, fed as a trace element in a mineral mixture or given as a drench, but all these methods have obvious disadvantages.

Perhaps the best preventive method is the twice yearly dosing of the individual sheep with the cobalt bullets.

7
Vitamin E Deficiency

THIS IS another chronic wasting disease of sheep which is also known as muscular dystrophy and white muscle disease *(photo 1)*.

Cause
The specific cause is a shortage of vitamin E, but the general predisposing cause is usually starvation.

Symptoms
Acute and rapid wasting of the skeletal muscles. The affected sheep is soon unable to stand. The heart muscle is also affected and death occurs within a few days from heart failure.

Treatment
A veterinary surgeon will be required to make the diagnosis and he may well have to seek the help of a scientist. Post-mortem examination shows a characteristic wasting and whitening of the muscle fibres on the skeleton and in the heart.

Provided the condition is not too far advanced the affected sheep respond to vitamin E given by injection and by the mouth. More important, however, is to prevent the rest of the flock going down with it.

Prevention
Improve the standard of nutrition immediately. Feed *ad lib* best quality hay and provide a concentrate mixture containing a vitamin supplement.

1

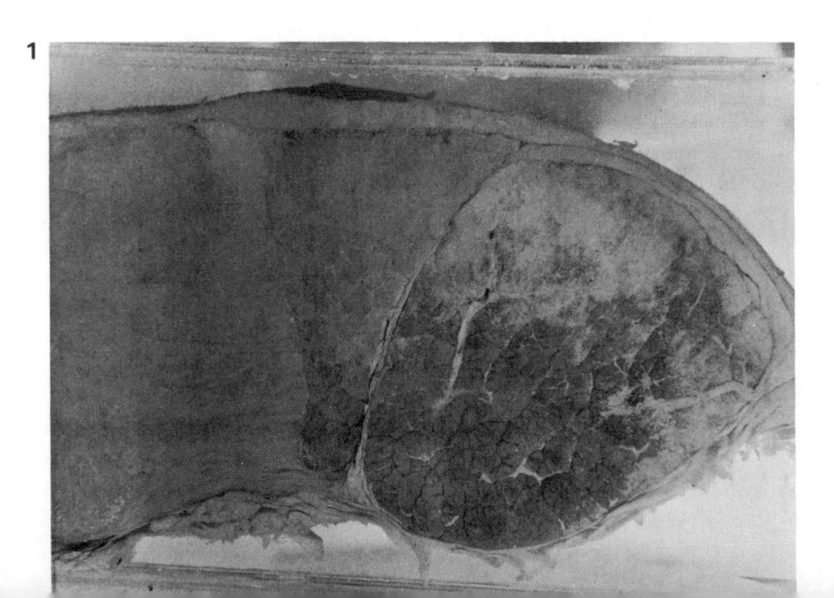

8

Phosphorus Deficiency

A MILD phosphorus deficiency in the pregnant or lactating ewe can produce a condition not unlike lambing sickness or mild hypomagnesaemia. In fact any cases of lambing sickness, or suspected hypomagnesaemia, that fail to respond completely to straight calcium or magnesium very often recover spectacularly when given combined calcium, phosphorus and magnesium *(photo 1)*.

However, severe and persistent phosphorus shortage can lead to a chronic wasting disease.

1

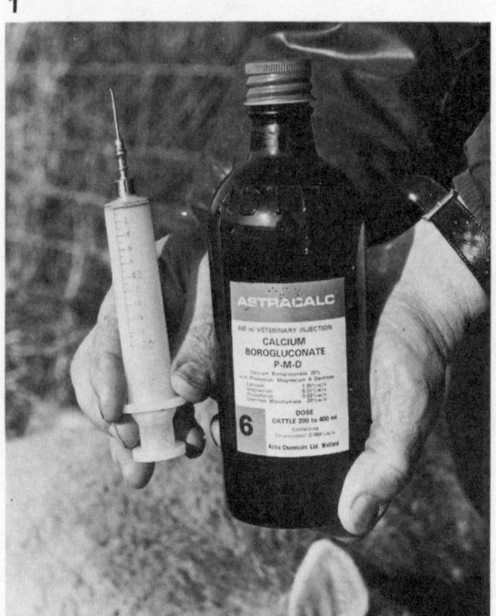

Cause

The usual cause in my experience is partial starvation on a phosphorus-deficient herbage.

Symptoms

The condition occurs in young growing sheep from 4 months old onwards. The bones do not develop properly, and this gives rise to three main conditions.

(a) DOUBLE SCALP

This is seen towards the back-end, especially on some of the poor hill pastures of Scotland.

Symptoms

The affected sheep are poor, light and unthrifty. The bones at the front of the skull are thin and flexible *(photo 2)*.

2

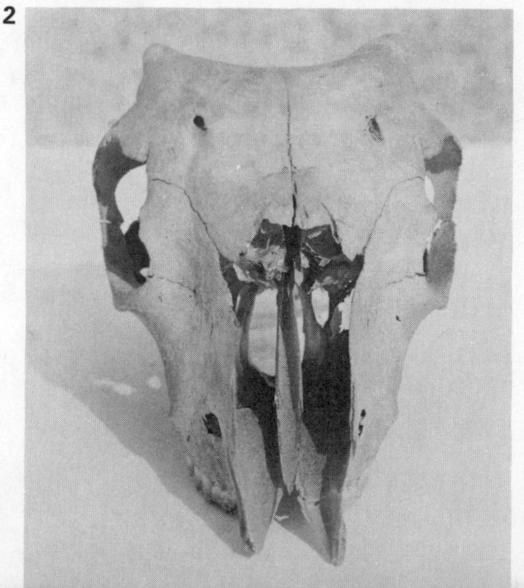

3

When you press these you can feel the inner layer of bone—hence the name, 'double scalp'.

Treatment And Prevention
It is not just enough to provide phosphorus. The overall plane of nutrition must be raised very considerably immediately. Change to a better pasture and supply best quality *ad lib* hay and concentrates. The affected sheep will then hold their own and will begin to recover more or less completely during the subsequent summer grazing.

Incidentally, since all unthrifty sheep are particularly susceptible to gastro-intestinal parasites, it is wise to inject or dose with anthelmintics before changing the pasture.

(b) RICKETS *(BENT-LEG)*

This is due to a deficiency of phosphorus and vitamin D, combined with rapid growth.

Symptoms
It is seen in young, rapidly growing lambs, especially when they are very well fed, for example, young ram lambs being fed for sale or when there is an excess of lush pasture in the spring. The rapidly growing lamb suddenly goes lame and one or several of the legs start to bend at the joints or in the shafts of the long bones *(photo 3)*. The affected joints may be swollen and painful.

Treatment And Prevention
Vitamin D injections, combined with calcium and phosphorus supplements during the time of rapid growth. A concentrate mixture containing 10 per cent of fish meal or 5 per cent of meat-and-bone meal will very often prevent further cases.

(c) OPEN MOUTH

Again, a disease of young growing sheep. It appears to be almost a combination of rickets and double scalp and is due to a complex of both these diseases.

Symptoms
The affected lambs cannot close their mouths and so the incisor teeth do not reach the dental pad. Naturally they rapidly lose condition because they cannot graze properly. The lower jaw bone is spongy and easily bent.

Treatment
Slaughter the affected animals.

Prevention
Concentrate on using the methods advised for double scalp and rickets.

CLOSTRIDIAL DISEASES

9

Enterotoxaemia

WHEN a sheep dies suddenly, the most likely cause of death is either hypomagnesaemia or enterotoxaemia.

Enterotoxaemia embraces three recognised conditions:

Enterotoxaemia of adult sheep.

Pulpy Kidney in lambs 3 to 12 weeks old.

Struck, seen in sheep over 1 year old.

Cause

A bacterium called the *Clostridium welchii*. There are two main types of this germ,

1

Type D and Type C. Type D produces the enterotoxaemia of older sheep and the so-called pulpy kidney of lambs. Type C causes struck in sheep over the age of 1 year.

When And Where Enterotoxaemia Occurs

Enterotoxaemia is an acute fatal disease occurring nearly everywhere and seen in sheep of all ages.

It produces pulpy kidney in lambs 3 to 12 weeks old, particularly single lambs.

It also hits weaned lambs and adult sheep, often affecting the biggest and best animals. In other words, it appears to hit the thriving lamb or sheep the hardest. The main danger period is the first few days after the flock is put on a changed or an improved diet, *i.e.* a rich pasture or a higher level of concentrates.

Clostridium welchii Type C hits sheep over 1 year old which are grazing high quality pastures in the late winter and early spring, but only in certain areas of Britain, *e.g.* Romney Marsh, North Wales and the Lothians.

Given favourable conditions, Types D and C grow rapidly in the intestines *(photo 1)* and excrete a highly lethal toxin. This toxin is absorbed and causes the death.

Symptoms

Sudden death is the rule though I have seen many sheep and lambs die from this

2

complaint. They fall down in a fit, throw their head back and kick furiously, apparently in abdominal pain *(photo 2)*. They then lapse into a coma and die.

There is no treatment.

Post-mortem Findings

It may interest veterinary student readers to know that I have found over the years that the presence of fluid (usually blood-stained) in the pericardial sac *(photo 3)* (i.e. the sac around the heart) is diagnostic of enterotoxaemia. This fact allows a fairly rapid confirmatory field diagnosis.

Prevention

There can be no excuse for any flockmaster who loses lambs or sheep from these enterotoxaemic diseases. If used correctly, the vaccines and antisera available are almost 100 per cent efficient.

However, for some reason which I shall never be able to understand, sheep farmers again and again forget or omit to vaccinate.

When an outbreak on such farms does occur, move the remaining sheep on to a bare pasture immediately or restrict their diet for several days before gradually building them back up to the original plane of nutrition.

In the meantime, having confirmed the diagnosis by consultation with the veterinary surgeon, protect the remaining flock with antisera pending the introduction of a full vaccination programme.

The compound vaccines are the best and most economical *(photo 4)*. Your veterinary surgeon will advise on the correct one to use to cover your particular disease problems.

Just one important point—booster vaccination immediately before lambing will give the lambs, via the colostrum, a protection for the first 12 weeks of life. If you are able to get your lambs off by that time, then there is no danger. However, all lambs kept over 3 months, whether as replacement breeders or fatteners, *must* be vaccinated. The ideal is to give them their first dose at 6 weeks and their second at 12 weeks.

3

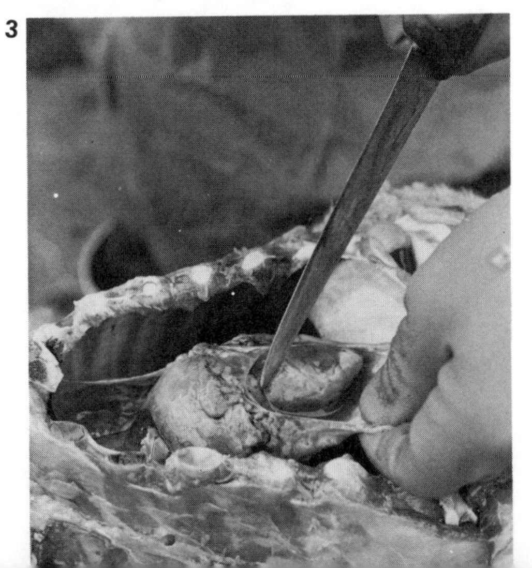

4

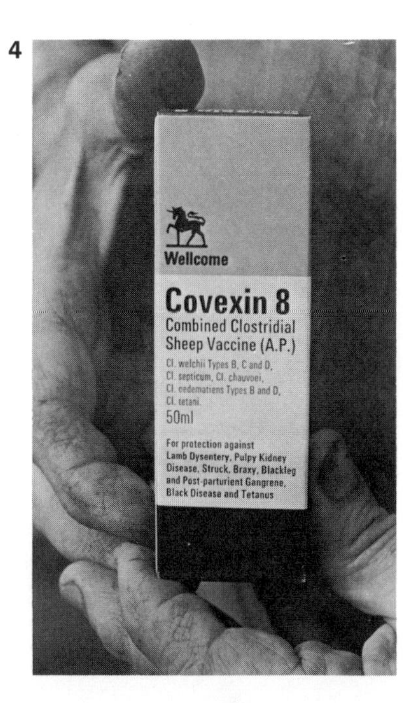

10
Lamb Dysentery

THIS IS without doubt the most dangerous of all sheep diseases. It attacks lambs under a week old though just occasionally it can knock them down at 2 or even 3 weeks *(photo 1)*.

Cause
It is caused by *Clostridium welchii* Type B. This germ produces its effect in the same way as Types D and C, that is, by multiplying in the small intestine of the lamb and excreting highly lethal toxins. The mortality rate among infected lambs is virtually 100 per cent.

It is a particular problem in the border counties between England and Scotland and in North Wales, but it can flare up anywhere.

Symptoms
The first sign noticed usually is the sudden death of one or two lambs. Closer observation will show that several others are dull and lethargic. They stop sucking and may start to kick at their belly. They stand with their backs arched and when made to move, they straddle rather than walk *(photo 2)*, then flop down. Within 24 hours, if still alive, they develop a profuse brown, often blood-stained diarrhoea *(photo 3)*. In the words of one of

1

2

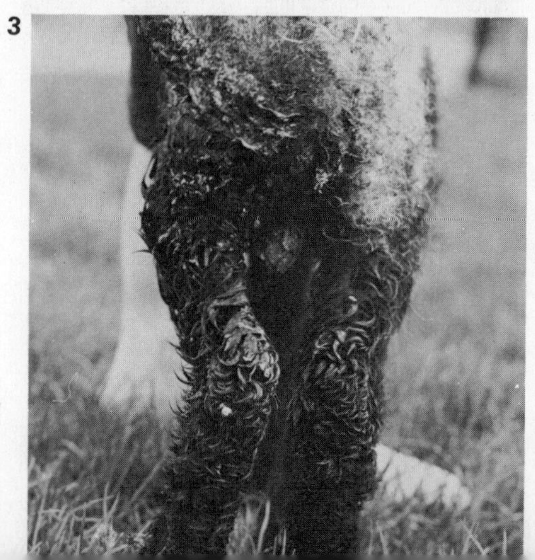
3

my shepherd friends, 'The poor little beggars strain their guts out.' They become dehydrated, comatose and die approximately 1 day after the onset of the diarrhoea.

In 2- to 3-week-old lambs the diarrhoea is less severe and the patients may survive for 2 or 3 days.

Treatment
Once the symptoms have developed, treatment is a waste of money and time, but each of the unaffected lambs should be protected immediately by injecting them with 2 cc of concentrated lamb dysentery serum *(photo 4)*. Any further new-born lambs during that lambing season must be given the serum as soon after birth as possible.

Prevention
Obviously this disease must be controlled, and it can be by vaccinating the ewes in all subsequent years with a multivalent sheep vaccine. Two doses the first year— 6 weeks and immediately before lambing and a booster dose immediately prior to each subsequent lambing.

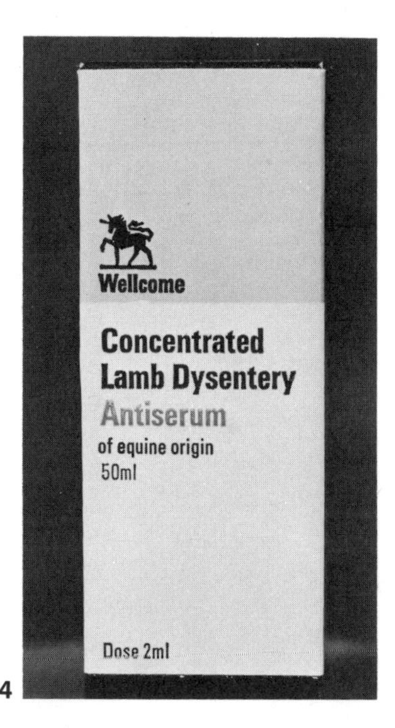

4

Any new breeding stock must have the double vaccination during the first year on the farm.

11
Braxy

THIS IS a very acute disease of young sheep occurring mainly in hill flocks between October and March.

Cause
It is caused by the *Clostridium septique (see diagram)* which also excretes lethal toxins; the growth of this germ occurs in the ewe's fourth stomach (the abomasum).

The *Clostridium septique* multiplies in the wall of the abomasum producing a powerful toxin, which is absorbed and causes a toxaemia.

Outbreaks usually coincide with the onset of frost.

Symptoms
It is usual to find several sheep dead. If seen alive, the braxy patient stands away from the rest of the flock. It is dull, refuses to eat, runs a high temperature of around 107° F and shows signs of abdominal pain, viz. grunting, grinding of the teeth and shifting the feet.

Within an hour or two it lies down; the temperature drops to below normal and death follows rapidly.

Clostridium septique with spores

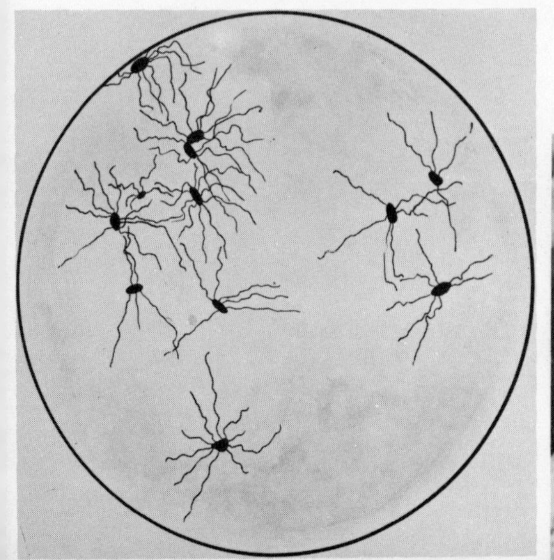

Post-mortem Signs
If the carcase can be opened up immediately after death, there is a clearly marked area of purple congestion or ulceration on the lining of the fourth stomach *(photo 1)*. Unfortunately this area disappears if the carcase is not examined within a few hours of death.

Treatment
There is none.

Prevention
Comprehensive vaccination, in all braxy areas, using an appropriate multivalent vaccine twice the first year and once each subsequent year during the early part of September.

1

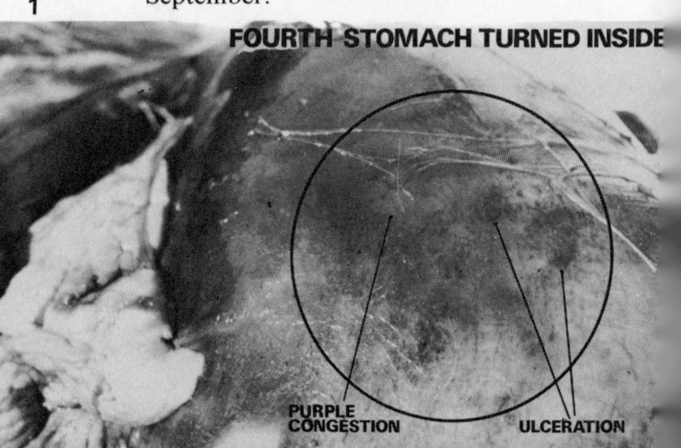

12
Black Disease

AN ACUTE fatal toxaemic disease affecting mainly adult sheep *(photo 1)*.

Cause
Another member of the Clostridial family —the *Clostridium oedematiens*, which multiplies in damaged liver tissue.

The trigger factor is fluke infestation. The liver damage caused by the invasion of immature flukes presents an ideal medium for the germ's growth *(photo 2)*. The bug proliferates rapidly and excretes lethal toxins which are absorbed into the bloodstream.

When And Where It Occurs
It is found on any hill or lowland farm which is infected by *Fasciola hepatica* (the sheep fluke).

Outbreaks coincide with the activity of the fluke parasite, that is, chiefly in the autumn and early winter and just occasionally in early spring.

1

2
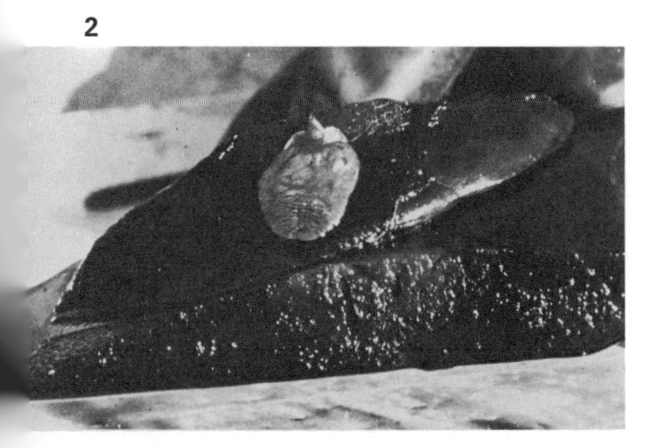

Adult sheep are the more susceptible, though I have seen outbreaks in lambs facing their first winter.

Symptoms
Again, comparatively sudden death is the rule since the course of the disease is a short one.

If seen in the early stages, the affected sheep lag behind the rest of the flock. When driven they collapse and breathe heavily. If left alone, they appear to fall asleep lying on their belly and brisket with their head round to one side *(photo 3)*. Any sudden noise will make them start violently and the eyes and ears twitch nervously. Shortly afterwards they go into a coma and die within a few hours.

Treatment

Treatment is useless in affected animals, but the remainder of the flock should be protected immediately with black disease antiserum. This gives only up to 20 days protection, so it is wise to give a first dose of black disease vaccine at the same time as the serum, and give a second booster dose a month later.

3

Prevention

Obviously this is another disease which should never be allowed to flare up. All sheep being kept in fluke areas must be protected by annual vaccination with a multivalent vaccine—twice the first year and a single booster each subsequent August or September.

13
Clostridial Infections of Wounds

THIS GENERAL heading incorporates three commonly recognised conditions:

(a) *Post-parturient Gas Gangrene.*
(b) *Blackquarter*—Blackleg *(photo 1)*.
(c) *Malignant Oedema (see diagram B)*.

Cause

Four clostridia are involved in post-parturient gas gangrene and in other wound infections:

Clostridium septique.
Clostridium welchii.
Clostridium chauvoei.
Clostridium oedematiens.

In blackquarter or blackleg, *Clostridium chauvoei* is the bacterium most frequently recovered *(diagram A)*. In so-called malignant oedema, the chief 'offender' is the *Clostridium septique.*

Diagram B—Malignant Oedema of a Muscle

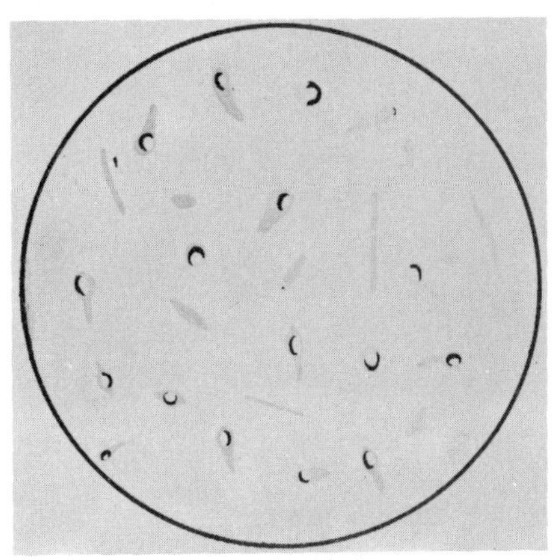

Diagram A—Blackleg bacilli

1

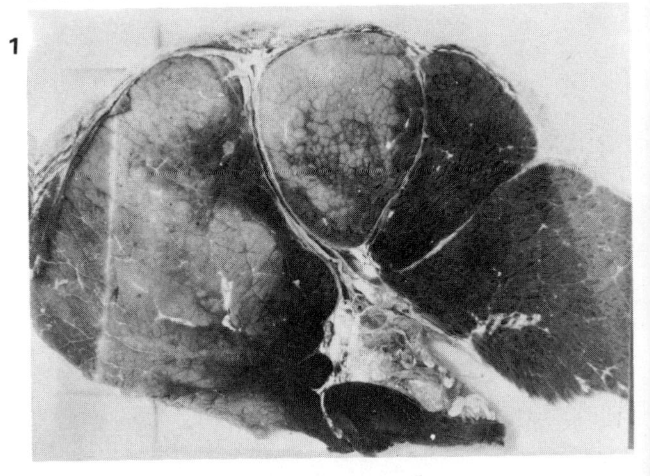

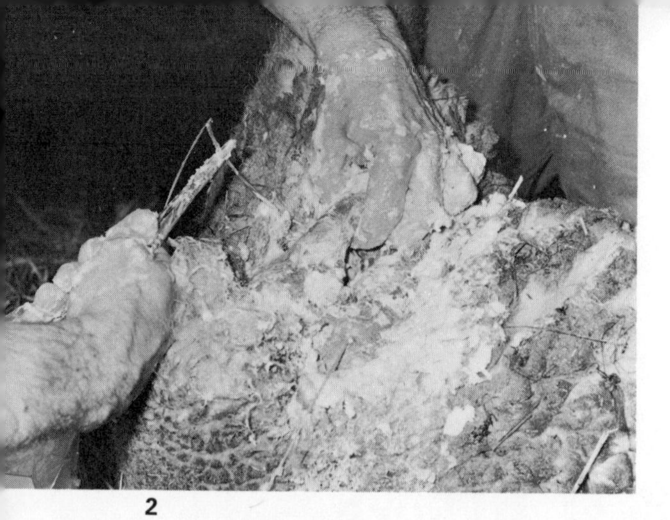

2

When Infection Is Likely To Occur
Obviously, infection is most likely when wounds are inflicted on the sheep *(photo 2)*. For example, during or after lambing, castration, docking, shearing and also head wounds caused by rams fighting.

Deep puncture wounds are the most dangerous. Unlike the blackleg of cattle, blackquarter in sheep flares up only as a result of wound infection.

Symptoms
Two days after the respective wounds become infected the sheep become depressed and dull and run a very high temperature (107° F). They walk with difficulty, the wound area is swollen *(photo 3)*, soft and hot but later becomes tense and hard. Pressure over the area usually shows crepitation, that is, when you run your fingers over the skin surface, it feels as though there was tissue paper underneath.

Treatment
The only hope is to catch the condition very early—then daily injections of penicillin may possibly save the sheep. Usually, however, cases are too far gone before being seen. Obviously, therefore, it is much better to do everything possible to prevent such infections.

Prevention
All these conditions can be controlled by the routine use of a comprehensive vaccine covering the four offending clostridia. Give two doses of the multivalent vaccine the first year, then a booster immediately before lambing, and another booster each year again just before lambing.

Despite this protection, it is always wise to use strict hygienic precautions at lambing time.

Also, a routine which I invariably advise after every bad lambing case—in fact whenever any assistance has to be given—is a single dose of long-acting penicillin injected preferably intramuscularly *(photo 4)*.

3

4

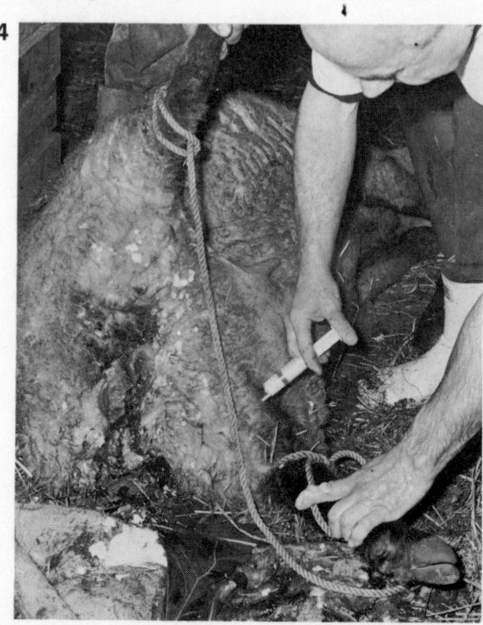

14
Tetanus

TETANUS IS primarily a disease of lambs, though all ages of sheep can be affected.

In the lambs it occurs 2 to 3 weeks after castration and docking, especially where the rubber ring method is used *(photo 1)*.

In ewes it may be seen shortly after lambing or after shearing.

As a rule, several animals are affected and, like braxy and black disease, tetanus tends to occur on the same farm year after year.

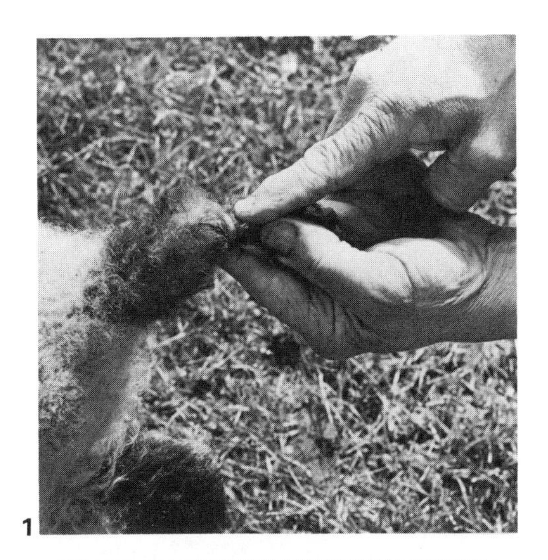

1

Cause

It is caused by the toxins produced by the germ the *Clostridium* or *Bacillus tetani (see diagram)*, which is commonly present in animal dung and in the soil.

The *Clostridium tetani* is a sporulating bacterium, *i.e.* it surrounds itself by a protective covering which enables it to live in the soil for many years.

Any wound, particularly behind or underneath the animal, is liable to become contaminated by the tetanus spores.

If the wound is deep, the bug casts off its protective coat and starts to multiply.

Two toxins are excreted and one of these—a neuro (or nerve toxin)—travels along the nerve fibres to the central nervous system.

The damage it causes in the central nervous system (the brain and spinal cord) results in increased irritability and acutely painful muscle spasms.

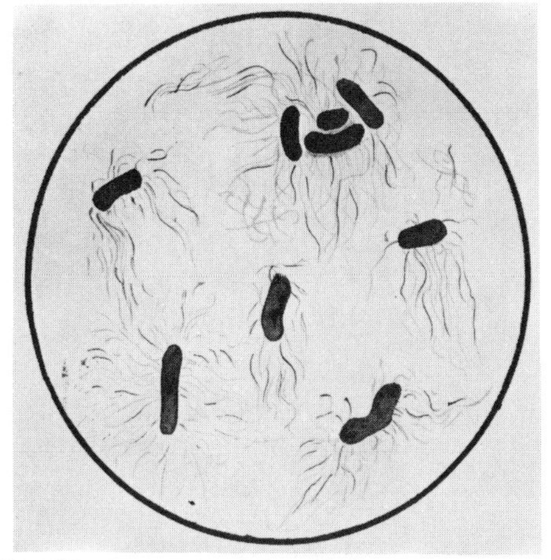

Bacillus tetani with spores in various stages of development

Symptoms

The symptoms may be mild or acute.

In mild cases the affected animal merely appears stiff and walks in a peculiar stilted fashion *(photo 2)*, just as though it were suffering from the after-effects of the castration, docking or lambing. Such mild cases may recover completely without treatment.

In acute cases the stiffness is much more obvious and widespread. The muscles are hard and rigid: the head and neck are extended *(photo 3)* and the tail cocked.

The patient is hypersensitive; it has great difficulty in lowering its head, and the jaws become locked. Eventually it falls over on its side and lies with legs outstretched and head thrown back.

Treatment

Mild cases appear to respond to daily injections of penicillin (6 cc of 300,000 units per cc).

Acute cases do not recover and should be destroyed.

Prevention

Comprehensive vaccination, coupled with strict aseptic precautions and general cleanliness during castration, docking, lambing and shearing, *i.e.* whenever a wound is likely to be produced.

2

3

15
Botulism

THIS IS an uncommon disease of sheep seen mainly in starving and scavenging animals.

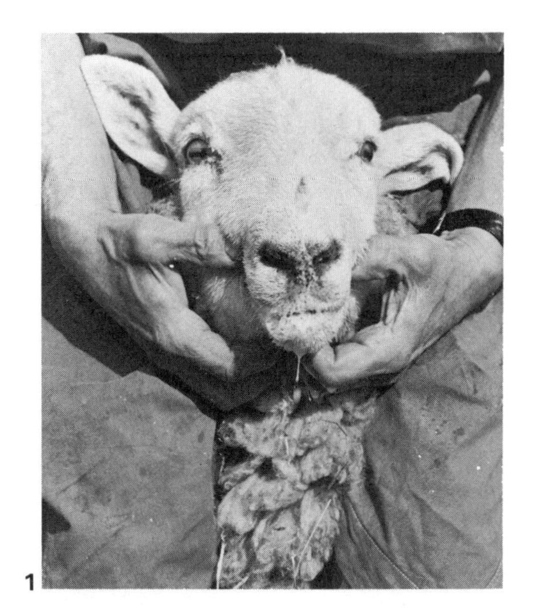
1

Cause
The toxin of *Clostridium botulinum.*

Source
Decomposing animal material. The starving scavengers are so short of protein and phosphorus that they will literally eat anything. When they swallow the *Clostridium botulinum*, the lethal toxin causes a progressive muscular paralysis.

Symptoms
The affected sheep salivates profusely *(photo 1)* and walks stiffly. Within a few hours it starts to bob and sway like a drunken person and has great difficulty in breathing. Eventually it flops down and dies, simply because the chest muscles become paralysed and it can no longer draw breath.

Treatment
There is none.

Prevention
Reasonable feeding—*never starve your sheep.*

DISEASES OF TICK AREAS

16

Louping Ill

(Trembling)

THIS IS a virus disease spread by ticks and affecting hill sheep in Scotland and Northern England. It breaks out during the months of greatest tick activity—April, May, early June and September.

When it appears in a flock for the first time, it affects sheep of all ages and can kill off up to 60 per cent. Once it is established, it recurs each year but much less virulently, attacking chiefly the yearlings and killing off no more than 4 per cent of the flock.

Symptoms

After an incubation period of 1 to 3 weeks, the virus invades the bloodstream causing a high fever. The affected animals are dull, they breathe heavily and are usually off their feed *(photo 1)*.

The fever subsides after a few days and many recover, but a week later typical nervous symptoms may develop. The patient becomes hyperexcitable and starts to tremble (hence the name, 'trembling'). The trembling is most marked around the head and neck. At this stage the temperature may rise again.

As the disease progresses, the patient may start to prance like a trotting pony or it may suddenly leap forwards and fall flat, kicking its legs as though in a fit.

Treatment

There is no satisfactory treatment. Seda-

1

tion by heavy doses of tranquilliser plus hope. If paralysis of the hind legs does not occur, the patient stands a chance of apparent recovery, though it may remain a carrier of the virus.

Prevention

Obviously it is vital to control the tick so far as possible. Regular dipping combined with heather burning and bracken cutting will assist greatly.

However, in all tick areas it is wise to combine such precautions with the rigid routine use of the louping ill vaccine *(photo 2)*. All sheep should be vaccinated twice the first year—in the early spring (March) and late summer (August)—and thereafter, if the flock remains self-contained, all ewe lambs should be given a booster dose each August and again in March.

Just one more interesting point. Louping ill is transmissible to man, causing a high fever and acute muscular pain, though fortunately it is rarely fatal in humans.

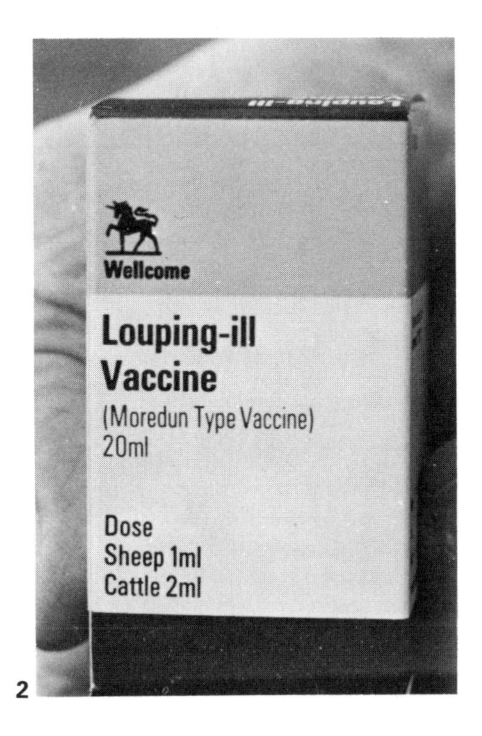

2

17
Tick-borne Fever

A NON-FATAL but debilitating fever which can attack sheep of all ages in the tick-infested areas of Scotland and Northern England. As with louping ill and tick pyaemia, it occurs when the ticks are most active in April, May, and early June, and again in September.

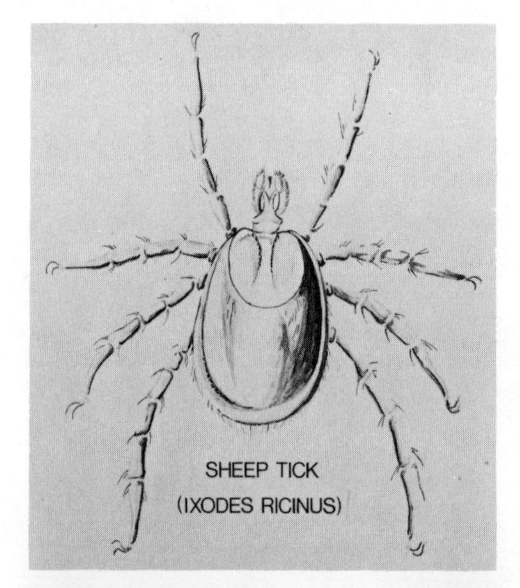

Cause
A rickettsial organism which is transmitted by the tick, *Ixodes ricinus (see diagram)*. The source of infection is the blood of an infected or of a carrier sheep.

Symptoms
The affected animals run a temperature of between 104° and 107° F for 9 or 10 days *(photo 1)*. The fever then subsides, but not before it has played havoc with the condition of the ewe or lamb and has lowered their resistance to other diseases. In this respect it is not unlike the glandular fever seen in humans.

Treatment
There is none.

Prevention
Maximum control of the ticks by regular dipping, bracken cutting and heather burning is, so far, the only defence against tick-borne fever, as the scientists have not yet succeeded in producing an efficient vaccine.

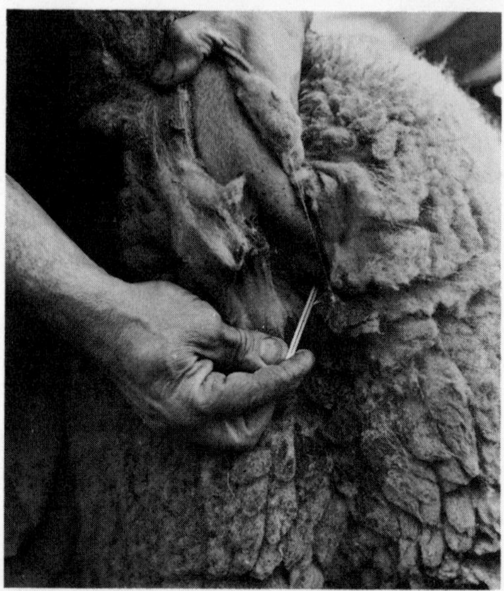
1

18
Tick Pyaemia

1

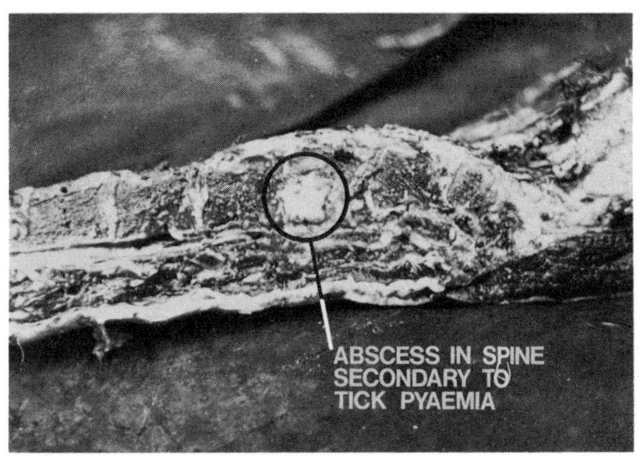

2

THIS IS a blood poisoning of lambs occurring in tick areas in April and May when the ticks are most active *(photo 1)*.

Cause
The same germ that causes most cases of mastitis in ewes, namely, *Staphylococcus aureus*. It gains entrance to the body through the tick bites, and if the lambs live long enough, it produces multiple abscesses throughout the body including joints, tendon sheaths, ribs, spinal column *(photo 2)*, brain, liver, spleen, kidneys and heart wall.

Symptoms
The acute cases run a high temperature *(photo 3)* and stop sucking. They are

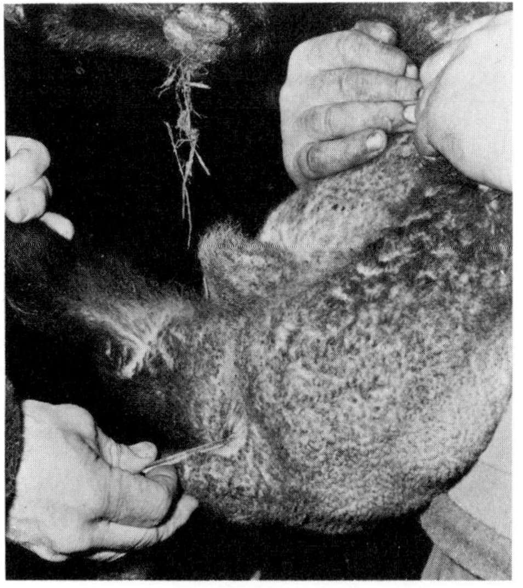

3

very lame and when they attempt to move show an inco-ordination not unlike that seen in louping ill. They go off their legs and die, and are usually infested with ticks.

The less acute cases may be indistinguishable from joint-ill showing lameness and swollen joints full of pus *(photo 4)*.

Treatment

Once affected, most lambs die or have to be destroyed and the mortality may be up to 20 per cent of the lamb flock.

If seen very early, a 5-day course of procaine penicillin (300,000 units daily) injections might effect a cure, but certainly advanced cases are much better destroyed.

4

Prevention

Again, commonsense control of the ticks is vital but here the obvious snag is that the lambs will not have been dipped.

Where tick pyaemia is a problem, the lambs should be dipped in a benzene hexachloride solution shortly after birth.

There is no satisfactory vaccine, and routine preventive use of antibiotic injections would be impracticable.

ADDITIONAL DISEASES OF LAMBS

19
Scour in Lambs

1

2

(a) E. Coli Infection

SINCE THE nightmare of lamb dysentery has been more or less completely subdued by general vaccination, another form of scour in lambs *(photo 1)* has moved in to take its place as the chief menace to lambs during the first week of life. This is *Escherichia coli* infection.

Cause

A bacterium called the *Escherichia coli*. There are several different strains of this germ.

The predisposing cause is lambing in a confined area. There appears to occur a build-up of infection as *E. coli* infection is usually most severe towards the end of lambing.

Symptoms

The affected lambs stop sucking and stand around with their backs arched *(photo 2)*. They develop a bright yellow diarrhoea and, if untreated, they die within a couple of days.

A peculiar feature of the disease is that it does not always appear to be acutely contagious, because one lamb of twins or triplets may be affected while the others remain perfectly normal.

Some of the affected lambs may become hyperexcitable and even blind.

Treatment

Get your veterinary surgeon to take swabs

of the faeces to have the *E. coli* typed and a sensitivity test done. He will provide an oral general antibiotic mixture *(photo 3)* as immediate protection pending the laboratory report. When the disease is confirmed, it will probably be necessary to dose all new-born lambs with the specific antibiotic.

Prevention
Move the ewes on to a fresh lambing area half-way through lambing or, better still, move them twice during the lambing period.

(b) Coccidiosis

This disease can produce diarrhoea in lambs 1 to 4 months old.

Cause
Certain species of coccidia.

Symptoms
The usual sign is dark blood-stained diarrhoea which soils the hind-quarters *(photo 4)*. If the lambs are in good condition, they recover fairly rapidly without treatment.

In other cases, however, the blood-stained diarrhoea persists and the lambs become tucked up and unthrifty.

Treatment
Call in your veterinary surgeon. He will take dung samples to identify and confirm the causal parasites and prescribe the appropriate specific treatment.

One of the best tried methods of treatment has been a 5-day course of sulphamezathine by the mouth, followed by a complete change of pasture.

Prevention
Move the lambs to fresh and extensive grazing as soon as possible.

From 6 to 12 weeks of age, nematodirus infestation is the commonest cause of scour in lambs (see Nematodirus, p. 70).

3

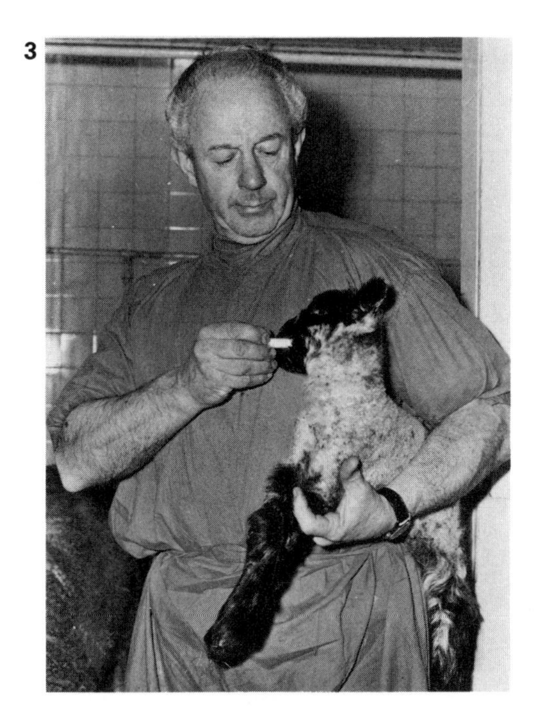

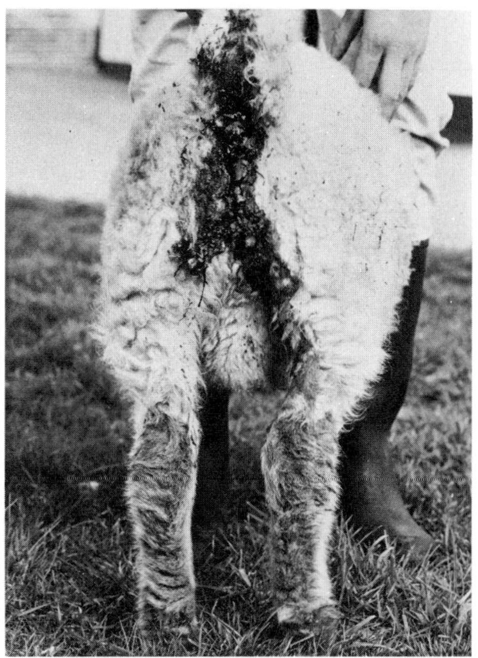

4

Thereafter, ordinary stomach and bowel worms are the main culprits (see Gastro-intestinal Helminthiasis, p. 66).

20
Daft Lamb Disease*

1

2

LIKE SWAYBACK, daft lamb disease is a condition of new-born lambs and is the only disease I know which is likely to be mistaken for swayback; in fact the two conditions are sometimes indistinguishable *(photo 1)*. Fortunately, however, daft lamb disease is uncommon and is seldom, if ever, a major problem.

* *Seen chiefly in the Border Leicester breed and its crosses.*

Cause
Daft lamb disease is *not* due to a mineral deficiency but is genetic in origin. Part of the lamb's brain—the cerebellum—is congenitally atrophied (imperfectly developed).

Symptoms
Most cases have a characteristic jerking backwards of the head. They carry the head high and have the mouth pointing

backwards *(photo 2)* or towards the side. In the more severe cases the lambs are blind and walk in circles *(photo 3)*.

Treatment

There is none, though mild cases can be hand reared and should be given a chance.

Most cases are destroyed by shepherds or die of exposure or starvation. Those that can walk may improve gradually, and some may become apparently normal except when excited.

Prevention

Cull the mothers and change the tup.

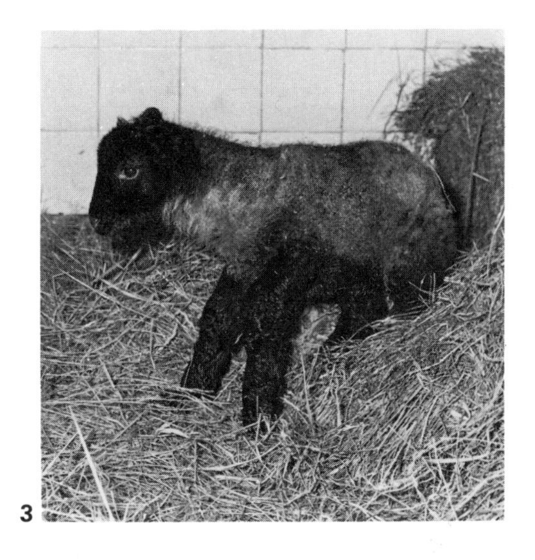

3

21
Joint-Ill

A DISEASE affecting the joints of young lambs and seen especially when the ewes are lambed in yards *(photo 1)* or in dirty pens in a field. In such conditions the disease incidence can be very high.

Cause

Various germs chiefly *Streptococci*, *Corynebacterium ovis* and *Bacterium coli*, all of which gain entry to the bloodstream via the umbilicus or navel cord shortly after birth.

1

2

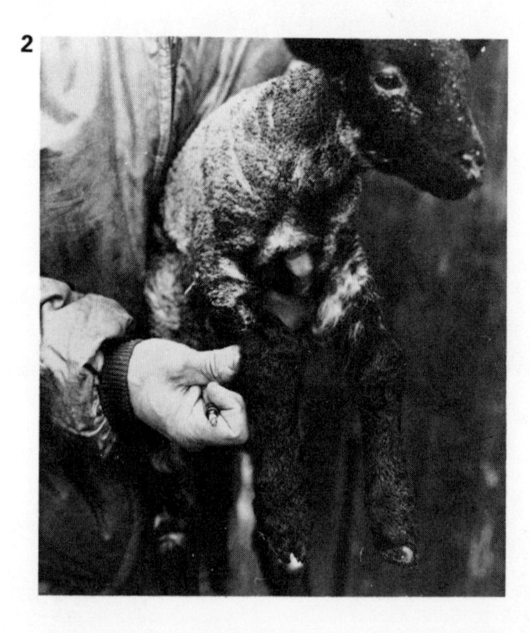

Symptoms

An initial fever with the temperature rising to 105°–106° F stops the lamb sucking. It becomes tucked up and then suddenly lame with one or more joints markedly swollen and painful *(photo 2)*. Occasionally the swelling extends down the tendons at the back of the affected leg.

Treatment

If caught early, *i.e.* as soon as they stop sucking and show the high temperature, most cases respond to daily injections (for 5 days) of penicillin and streptomycin

3

(photo 3). However, once the joints become infected the majority of the lambs die within a week, while the few that recover do so slowly and remain lame and unthrifty.

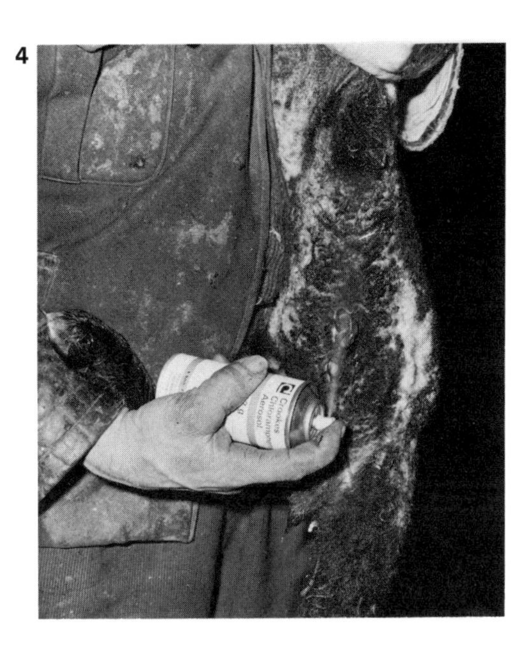

4

Prevention

Obviously this is a disease that shepherds should do everything possible to avoid, particularly since there is no satisfactory vaccine against it.

The accent must be on hygiene and general cleanliness.

Lambing pens should be bedded down with clean straw between each patient and the lambs' navels should be dressed as soon after birth as possible *(photo 4)*. The most convenient and perhaps the best dressing is an aerosol spray containing chloramphenicol, terramycin or aureomycin. If these are not available, a simple antiseptic dressing such as 10 per cent copper sulphate solution will help considerably.

Shepherds should use abundant water, soap, soap-flakes and antiseptic whenever assistance is necessary, and should dip their feet in antiseptic solution before entering the lambing pens.

22

Liver Abscesses

(Hepatic Necrobacillosis—Fusiformis Necrophorus Infection)

THIS CONDITION can occur at anytime on most lowland farms. It affects lambs 5 to 7 days old *(photo 1)* and can kill off 5 to 10 per cent of the lamb flock.

1

Cause

It is caused by a germ called the *Fusiformis necrophorus*—the same germ that is involved in foot rot (see Foot Rot, p. 80). The Fusiformis gains entrance through the navel in new-born lambs.

Symptoms

One or several lambs that have been doing really well for the first few days suddenly stop sucking and literally fade away. The abdomen often swells up and death follows in a few days.

The disease is confirmed by post-mortem examination, when numerous white abscesses or caseating vile-smelling dead nodules are found in the liver *(photo 2)* and occasionally in the lungs also.

Treatment

As in joint ill, treatment has to be started very early, *i.e.* just as soon as the lamb stops sucking. Then a 3-day course of penicillin injections *(photo 3)* will often save the patient (approximately 300,000 units of penicillin per day).

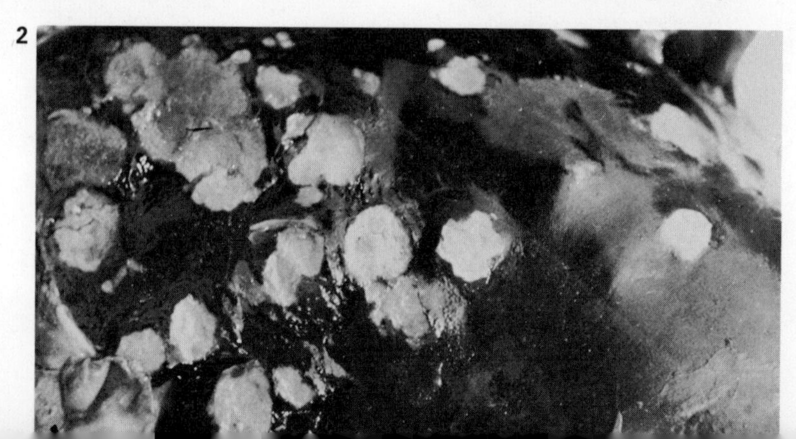

2

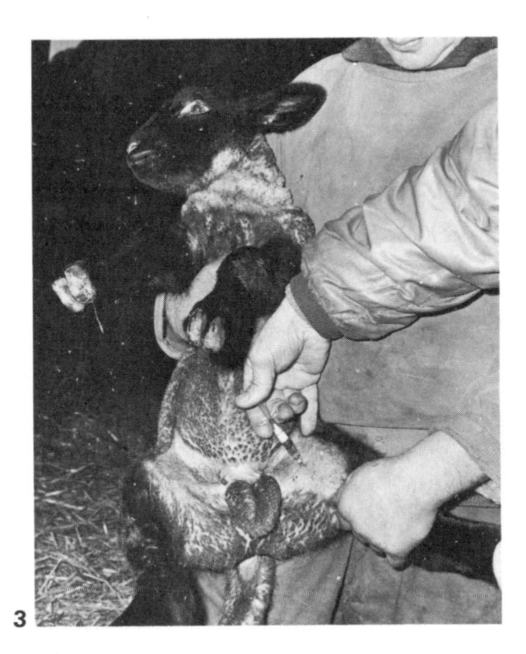

3

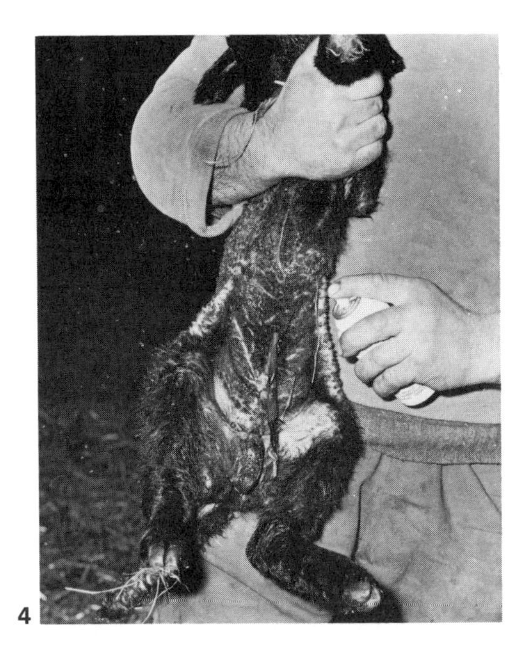

4

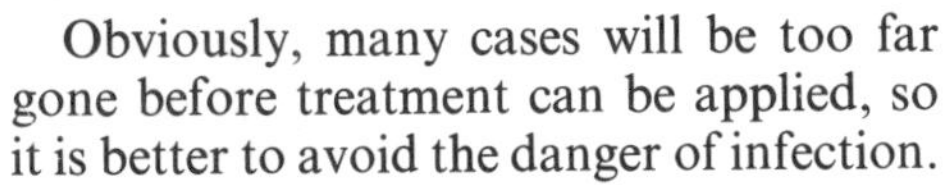

Obviously, many cases will be too far gone before treatment can be applied, so it is better to avoid the danger of infection.

Prevention

The advent of the foot rot vaccine should do much to control this condition. The entire ewe flock should be fully protected at all times. In this way the danger of the lambing pens becoming contaminated will be reduced to a minimum.

The same hygienic precautions as advised for joint ill should be strictly imposed, including the spraying of the navels of the new-born lambs with a broad spectrum antibiotic *(photo 4)*.

Where the lambing is being done in the open pasture, a change of field will often prevent new cases developing.

23
Stiff Lamb Disease

A DISEASE of growing lambs, 3 to 4 months old. The highest incidence occurs in East Anglia on arable land.

Cause
A germ called the *Erysipelothrix rhusiopathiae*, *i.e.* the same germ that causes swine erysipelas.

The erysipelothrix bug is believed to come from the soil and to get into the lambs through docking or castration or other wounds. It may also be eaten by the lambs *(photo 1)*.

Rarely do more than 5 per cent of an infected flock show signs of the disease, the onset of which is gradual.

1

2

What usually happens is that one or several lambs may be seen to be slightly lame; later a few more may be noticed. Rarely is there any history of contact with pigs.

Symptoms
The affected lamb is lame, but there is usually no obvious swelling of the joints. The lamb merely walks stiffly—hence the name, 'stiff lamb disease' *(photo 2)*.

The lameness persists for a long time and the patients do not thrive, simply because the *Erysipelothrix rhusiopathiae* produces a painful deep-seated arthritis.

Treatment

In early cases, penicillin injections will destroy the bugs and prevent a chronic arthritis. Once the condition has become well established, however, treatment has little chance of success. Obviously, therefore, as with so many other of the sheep diseases, it is much better to do everything possible to avoid stiff lamb disease.

Prevention

Strict asepsis at docking and castration. The lambs should be dropped on a pile of clean straw and then turned on to a pasture immediately after the operation.

They should be kept at grass until the wounds are healed. In other words, *never turn newly docked or castrated lambs on to arable land.*

I have used swine erysipelas vaccine with some success *(photo 4)*, injecting the ewes with 2 cc at the same time as the multivalent clostridial vaccines. But the disease can and should be controlled by simple cleanliness and commonsense.

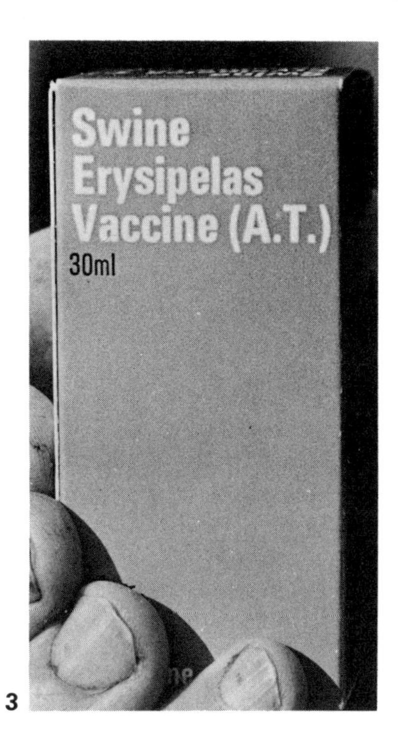

3

INTERNAL PARASITES

24

Roundworms

(Gastro-Intestinal Helminthiasis)

ROUNDWORMS IN sheep *(photo 1)* still produce greater losses to the sheep industry than any other single cause or disease, despite the free availability of the many excellent modern anthelmintics.

In my opinion this is due to relaxation in rigid routine dosing, combined with the intensification of stocking associated with improved grassland management. Productive pasture leads to heavy stocking and heavy stocking leads to excess parasites.

In order to understand this fully and to appreciate the necessity of good pasture husbandry and regular routine dosing, a knowledge of the life cycle of the worms is absolutely vital.

1

LIFE CYCLE OF THE SHEEP ROUNDWORMS

Approximately ten different species of roundworms live in the stomach and intestines of British sheep and all of them basically have the same life cycle *(see diagram)*.

Eggs, laid by the adult female worms inside the stomach or bowels, are passed out with the sheep's droppings on to the pasture.

In 24 hours the eggs change into first-stage larvae which feed on bacteria and moisture. Within a short time the first-stage larvae change, by casting off their outer coat, into second-stage larvae. These

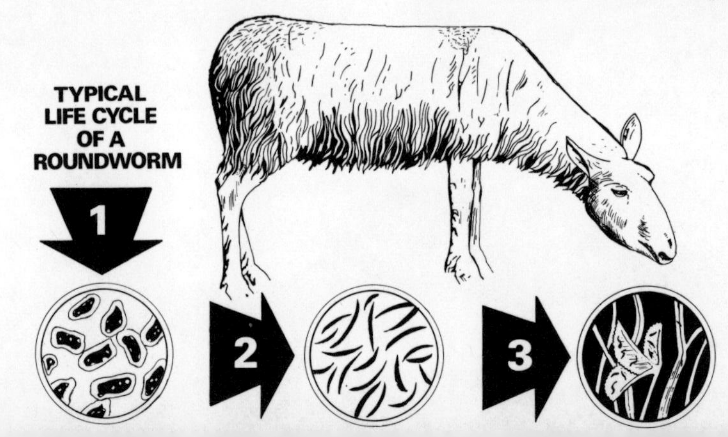

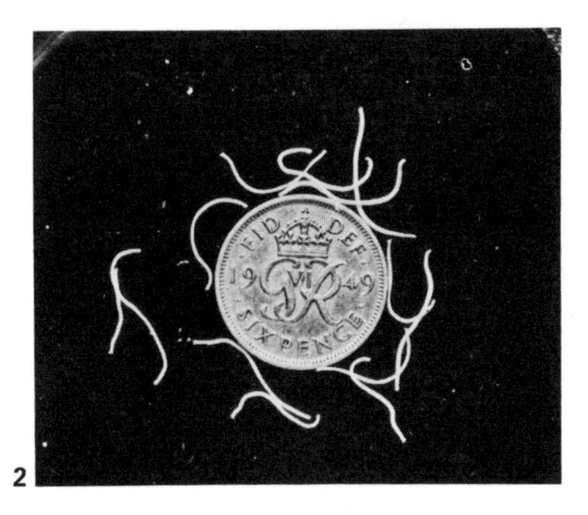

2

also feed on bacteria and moisture, then moult to form the third or infective stage. The infective stage retains the protective skin of the second stage. It develops in from 3 to 7 days.

The infective larvae, so-called because they are now ready to infect other sheep, cannot feed but they climb up and down the herbage waiting to be eaten by a grazing animal.

When they are swallowed, the infective larvae fix themselves to the lining of the stomach or intestines, and during the course of 3 or 4 weeks grow into adult male and female worms *(photo 2)* ready to copulate and produce another crop of eggs.

Symptoms

Lambs will tolerate a few parasites, but

3

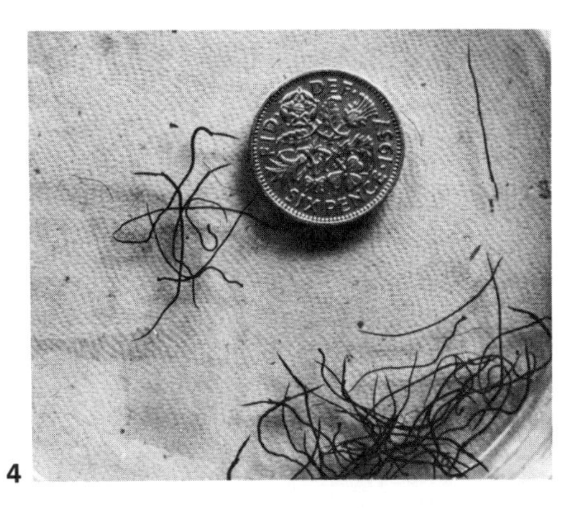

4

when large numbers are present severe symptoms, and very often death, result.

The symptoms depend on the species of worms that predominate, the age of the sheep, and the state of its nutrition.

When large numbers of worms, *e.g.* the trichostrongyles, are present in the intestines, diarrhoea or scouring is the outstanding symptom *(photo 3)*. The bowel becomes inflamed, the digestive products of the food are not absorbed and the affected animal becomes listless and stops thriving.

When stomach worms are present, *e.g.* the haemonchus *(photo 4)*, the ostertagia, etc., the most common symptom is anaemia. The worms attach themselves to the lining of the stomach to suck blood and cause gastritis and ulceration. Advanced cases develop a watery swelling under the jaw (bottle jaw), and the membranes of the eye become very pale, thus producing the 'white eye' condition reported by shepherds.

When, as often happens, both stomach and intestinal worms are present the symptoms are a combination of scouring, anaemia and rapid loss of condition.

Lambs are more seriously affected than adult sheep because the adults do possess a certain degree of immunity. As the lambs grow older, they also develop an immunity but this is never absolute and can be broken down by a gross infestation or by other stress factors. For example,

5

parasitic disease can flare up in adult sheep in a poor nutritional state or in ewes in advanced pregnancy.

The length of survival of the infective larvae in a pasture is obviously very important. There is no general agreement but it is known for certain that an infestation will remain alive for at least 3 months.

Prevention And Control

It is obvious from the life cycle that clean sheep can only become infected by eating the infective larvae from the pasture. Maximum contamination of the pasture occurs in the spring due to the so-called 'spring rise' in the output of eggs. At that time an apparently healthy sheep may pass out over eight million eggs daily.

The basis of control, therefore, must be to avoid a build-up of infective larvae on the pastures.

The first job is to kill the adult worms in the sheep. Using one of the modern anthelmintic injections, routine dosing should be started in the spring, about a fortnight before the later lambing *(photo 5)*. This will reduce the 'spring egg rise' to a minimum.

The lambs should be injected at 6 weeks and once every 6 weeks until the autumn, when the entire flock should be dosed again.

The dosing should be combined with sensible pasture husbandry. If the sheep are folded on small areas, for example, the hurdles should be moved every week at least since it takes 3 to 7 days for the larvae to become infective. Where ample grazing is available, the grazing should be changed after each dosing.

Ploughing-in will not kill all the larvae, but if another crop is taken off the land before grass is re-sown the land should be clear.

One sensible way of cleaning up pastures is by cross grazing them with cattle or horses; the sheep larvae cannot develop in these animals and are destroyed.

So far no pasture dressing has been found that will kill parasitic larvae without destroying the herbage.

LOWLAND SHEEP

A sensible routine for lowland sheep would be:

1. Dose the ewes in April or May when they go from the lambing field to the pasture. At this time the worms which make up the spring rise will be highly susceptible to anthelmintics.
2. Dose the lambs and ewes in mid-June and move to clean grazing. In early-lambing flocks this may coincide with weaning. If this is the case, the ewes need not be dosed.
3. Dose the lambs in late August and move to clean pastures. This may be the original pasture if it has been stock-free and hayed or silaged during the summer.

HILL SHEEP

Between October and April the only stock are the ewes; consequently they and the returning hoggs are the reservoirs of the worm eggs.

The spring egg rise in hill sheep occurs just before lambing. Therefore hill ewes should receive their first dose of anthelmintic about a fortnight before lambing.

Because of the growth of the grass after lambing and the consequent improved nutrition of the hill ewes during the summer, and also because of their wide grazing area, they will not require dosing again until the autumn.

The hoggs returning from the wintering in April should be dosed before being put out to the hill and again, as gimmers, with the stock ewes, in September.

If scouring develops in the lambs at any time, they should be collected and injected immediately.

25

Nematodirus Infection

NEMATODIRUS INFECTION attacks young lambs suddenly and, if untreated, kills them off rapidly.

Cause

A species of roundworm called the Nematodirus which is, without doubt, the most dangerous of all the roundworms *(photo 1)*.

It attacks in the spring—in March and April in Northern Ireland, and in May and June in the rest of Britain and especially in the Scottish border counties.

How Nematodirus Differs From Ordinary Roundworms

The fundamental practical difference between ordinary roundworms and Nematodirus is that the latter is destroyed only by the more powerful anthelmintics.

Also, it produces a disease which differs from the typical roundworm infestation in that it is sudden in onset, hyperacute, and rapidly fatal if untreated.

1

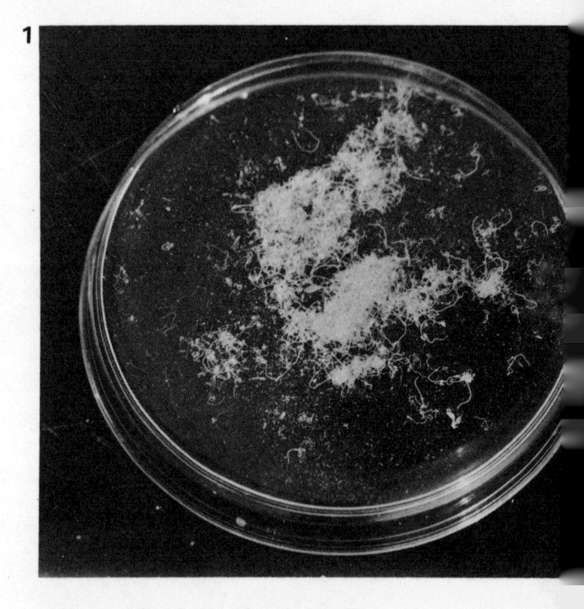

The disease occurs early in the year, *i.e.* before the traditional roundworm trouble

Worm Development Stage by Stage

WORM DEVELOPMENT STAGE BY STAGE

1. TYPICAL ROUNDWORM LIFE HISTORY
DEVELOPMENT TO EGG LAYING STAGE IN 3 WEEKS

SHEEP

3RD STAGE LARVAE (INFECTIVE)

PASTURE STAGES

EGGS DROPPED ON PASTURE

2ND STAGE LARVAE

1ST STAGE LARVAE FROM EGGS

DEVELOPMENT FROM EGG TO INFECTIVE 3RD STAGE LARVA IN MINIMUM OF 5 DAYS

2. NEMATODIRUS LIFE HISTORY
DEVELOPMENT TO EGG LAYING ADULT STAGE IN 2-3 WEEKS

SHEEP

INFECTIVE RD STAGE LARVAE (HATCHING IN SPRING AFTER 8-9 MONTHS INACTIVITY)

PASTURE STAGE

EGGS DROPPED ON PASTURE

LARVAE DEVELOPS WITHIN EGGS

3RD STAGE 2ND STAGE 1ST STAGE

DEVELOPMENT FROM EGG TO UNHATCHED 3RD STAGE LARVA IN MINIMUM OF 2-3 MONTHS

would normally be expected, and it affects only young lambs from 1 to 5 months of age.

Another peculiar feature is that the disease appears on pastures which have been rested over the winter and which, so far as the usual roundworms are concerned, would normally be considered clean and fairly safe.

What happens is shown in the diagram. With the typical roundworm the worm eggs are laid inside the sheep and are passed out in the droppings on to the pastures. There, in the course of less than a week, they undergo three changes—into the first-stage larvae, then into the second-stage larvae, and finally into the third or infective stage.

Many of the eggs are destroyed by climatic conditions but a considerable number reach the infective stage and, if the weather is mild, they can *keep a pasture infected for up to 6 months*. However, if there is a hot dry summer or an exceptionally cold winter, the larvae may survive for less than 3 months.

With Nematodirus the eggs pass on to the pasture in the same way but thereafter the *first-, second- and third-stage larvae develop inside the eggs and this development takes approximately 2 to 3 months*. The embryonated eggs of Nematodirus have great resistance against sunlight, drought and frost and they remain dormant throughout the entire winter to hatch out simultaneously the following spring.

The masses of newly hatched larvae are eaten by the susceptible young lambs, and inside the lambs the larvae moult and pierce the bowel lining.

Symptoms

Sudden onset of acute diarrhoea in a group of healthy spring lambs. The affected ones quickly become dehydrated; they make for water, drink it to excess and stand in it. They lose weight rapidly and die in 2 to 4 days.

Treatment

The affected lambs will respond to the correct maximum dose of one of the new powerful specific anthelmintics, but the most important thing of all is to spot the condition early and get to work immediately. Any delay can cost the loss of up to half the lamb flock.

Prevention

On farms where Nematodirus is known to exist, it is probably best to anticipate the trouble by preventative dosing at the beginning and end of the first danger month (*i.e.* May in most parts) with perhaps a third dose at the end of June. This triple dosing may be expensive but nothing like as expensive as an active infection could prove.

With the continual improvement in the modern anthelmintics, a single early spring dosing may be sufficient to prevent the majority of outbreaks.

Another useful preventative hint is not to use the same pasture for lambs in successive years.

26

Fluke: Fasciolasis: Liver Rot

(Liver Fluke Disease)

1

THIS IS an acute or chronic disease affecting sheep of all ages *(photo 1)*. It is more prevalent in some areas than in others and usually flares up after a warm wet summer.

2

Cause
It is caused by the liver fluke—called the *Fasciola hepatica (photo 2)*. In order to understand the development and control of the disease a knowledge of the life history of the fluke is essential.

Life Cycle
For easy reference I append the life cycle in simple diagrammatic form as well as in descriptive terms *(see diagram A)*.

Diagram A

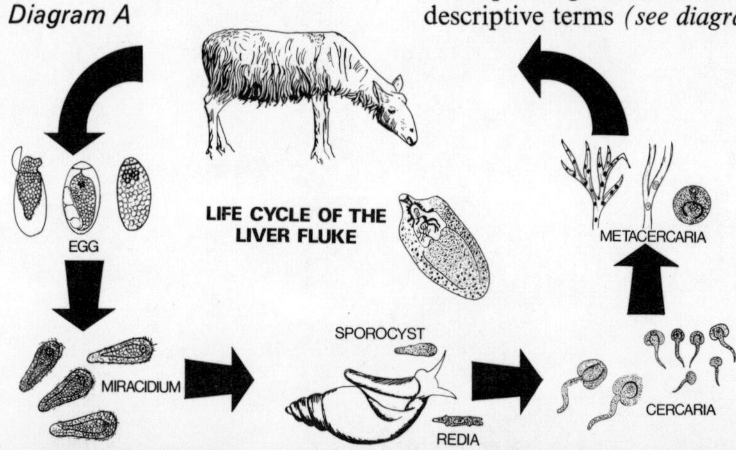

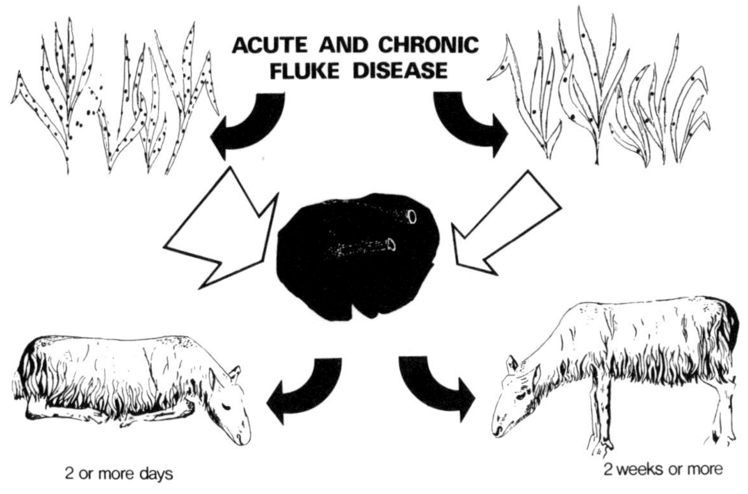

Diagram B

MATURE ADULT FLUKES — In the bile ducts of an infected sheep the mature adult flukes, which are hermaphrodite (bi-sexual), lay eggs. These are passed down the bile ducts into the intestine and out on to the pastures in the dung.

EGGS — On the pastures the eggs develop into the first-stage larvae—called the miracidia. This process can take place in 9 days or it can be delayed, in unsuitable conditions, up to 5 months.

9 DAYS — UP TO 5 MONTHS

The miracidia swim about and are picked up by a snail called the *Limnaea truncatula*.

SNAIL (*LIMNAEA TRUNCATULA*) — Inside the snail the miracidia undergo changes into sporocysts, then rediae and finally develop into the second-stage larvae called the cercariae. This process takes 6 to 7 weeks.

6 TO 7 WEEKS

CERCARIAE — The cercariae then leave the snail and attach themselves to the herbage where they are sometimes described as metacercariae. The cercariae are the infective larvae. They may be eaten immediately by another sheep or they may survive for up to 12 months or longer.

UP TO 12 MONTHS

SHEEP INTESTINE — When eaten by another sheep, the cercariae penetrate the intestinal wall and in the course of 4 or 5 days, they reach the liver *(see diagram B)*.

4 TO 5 DAYS

SHEEP LIVER — PARENCHYMA — The cercariae migrate through the substance of the liver—called the parenchyma (liver substance)—for 5 or 6 weeks, gradually increasing in size and of course damaging the liver in the process.

5 TO 6 WEEKS

BILE DUCTS — After that time, they enter the bile ducts and grow into mature adults in between 6 and 7 weeks.

6 TO 7 WEEKS

MATURE ADULT FLUKES — In the bile ducts the adults again excrete their eggs, and the life cycle starts all over again.

When established in the bile ducts, adult flukes will rarely attempt to leave. Some of them have been known to survive in the same sheep for 11 years.

In the words of one of our most famous parasitologists, Dr E. L. Taylor: 'In the final host the liver fluke lives a simple life with little change. Bathed continuously in a stream of bile, at a midsummer temperature that scarcely varies a degree from one end of its life to the other, it withdraws the good nutriment that is unfailingly provided. There are no enemies to molest it; nor does it have to look for a mate to reproduce its kind, but it continues to bring forth an amazing stream of innumerable eggs in order to ensure that the species *Fasciola hepatica* shall continue on the earth.' Such is the parasite you have to destroy and fortunately several anthelmintics will do just that.

LIFE CYCLE OF LIMNAEA TRUNCATULA

In order to understand the disease still further it is wise to study the life cycle of the *Limnaea truncatula*.

The snails, male and female, copulate in the early spring and the females lay their eggs in March or April.

The eggs develop into adult snails in 3 months, and these in turn produce a further generation in 3 months.

Since snails live in mud and since the miracidia have to swim about to find them, it is obvious that flukes can only survive and thrive on wet land. For example, peat bogs and water springs on hill land, badly drained land anywhere, open drains, ditches and low-lying field flanking a river which floods easily—in fact any wet or muddy land.

Symptoms

There are three types of the disease—hyperacute, acute and chronic.

In the *hyperacute* cases, sudden death is the rule and the condition is only diagnosed on post-mortem examination.

Acute fasciolasis does not begin until July and tends to get worse right up to November and December.

Several sheep, often lambs, become suddenly ill. They stand about, go off their food and are unwilling to move. When driven they fall down and refuse to rise.

3

The belly is swollen and painful, especially over the liver, and death takes place within a day or two.

Chronic fasciolasis is seen in the latter part of the winter or early spring. It is a wasting disease which is difficult to distinguish from roundworm infestation.

Usually several sheep—adults and lambs—stop thriving. When examined, they are very thin and light and often show an oedema or dropsy underneath the jaw *(photo 3)*. The mucous membranes of the mouth and eye are pale and anaemic.

Diagnosis can only be confirmed by getting your veterinary surgeon to examine the dung. Presence of the typical fluke egg or eggs will establish fluke infestation absolutely.

Treatment

Where hyperacute or acute fasciolasis breaks out, treatment is not spectacularly successful simply because even the modern highly efficient anthelmintics have little effect on young flukes under 6 weeks of age; they act only on the older flukes.

The entire flock should be dosed immediately with one of the modern fluke injections such as TRODAX (May & Baker), ZANIL (ICI) or COOPAPHENE (Coopers).

The treatment should be repeated every 3 weeks until 6 weeks has elapsed after the last fluke death.

Immediately after the final dosing remove the flock to a fluke-free area—such as a dry new pasture or aftermath: or to a ploughed field.

In less acute attacks it may only be necessary to repeat the initial treatment twice—at 4 and 8 weeks.

In chronic fluke outbreaks, treatment is highly successful. A single dose of any of the drugs mentioned will remove the fluke population from the flock, and the anaemia will cure itself in 2 or 3 weeks. After dosing, of course, it is wise to move the flock on to a dry clean area.

Prevention

On farms that have a fluke problem, control measures should be carried out during the winter months. If the summer has been warm and wet, then the entire flock should be injected once a month from October to a month or 6 weeks before lambing. This monthly dosing will destroy each batch of adult flukes shortly after they arrive in the bile duct, before they have had time to do a great deal of damage to the sheep's health, and more important perhaps, before they reach full sexual maturity. This will prevent further contamination of the ground since the immature adults will not produce any eggs.

Another way of tackling the problem is to reduce the snail population, and the best long-term way of doing this is by drainage of the danger areas.

Where the snail areas are comparatively small, they can be fenced off and the snails destroyed with a molluscicide (a snail killer). For many years copper sulphate was the only effective dressing, but nowadays more efficient drugs are becoming available, *e.g.* FRESCON made by Shell Chemicals Ltd.

The best time to apply the molluscicide is in the early spring—March—before the snail breeding gets going, and again, if necessary, in June before any infected snails can shed their cercariae.

The ideal, of course, in bad fluke areas is to do all three, viz.—drain, dress with a molluscicide, and dose with anthelmintics.

27
Tapeworms

TAPEWORM INFESTATION in lambs, although not common, can occasionally flare up and it is important to know how to deal with it.

Cause
Tapeworms belonging to the family or genus called Monieza *(photo 1)*. The worms are wide with short segments.

The segments of the adult worm are passed out in the dung. The cover of the segment disintegrates exposing the eggs. The eggs are eaten by common pasture mites, and inside these mites they grow into infective larvae.

The lamb eats the mites, along with the grasses, and in the lamb's intestines the tapeworm larvae break free and develop into adult tapeworms.

Symptoms
There may be a general loss of weight and unthriftiness, but usually the segments are either seen in the dung, or a lamb may die suddenly and be found to have a large number of tapeworms in the intestine.

It becomes a flock problem only occasionally.

Treatment
The best treatment for tapeworms in sheep is still one of the oldest of worm remedies, *viz.* copper sulphate and nicotine sulphate given by the mouth *(photo 2)*.

1

2

A solution should be made up as follows:

Copper sulphate crystals—1 lb 4 oz.
Nicotine sulphate (40 per cent solution)—1 pint.
Soft water—$2\frac{1}{2}$ gallons.
Hydrochloric acid—10 to 12 drops.

The doses are:
Ewes—1 oz.
Lambs 6 to 9 months—$\frac{3}{4}$ oz.
Lambs 3 to 6 months—$\frac{1}{2}$ oz.
Lambs 2 to 3 months—$\frac{1}{4}$ oz.

Prevention
If a heavy infestation is present, plough up the pastures or cross graze with cattle for a full season.

FOOT CONDITIONS

28
Foot Rot

THE CHIEF causal organism is a germ called the *Fusiformis nodosus*. This microbe cannot survive outside the sheep's foot for more than 9 or 10 days, nor can it multiply in the soil. Obviously, therefore, the *only* persistent source of infection in any flock is a sheep suffering from foot rot *(photo 1)*.

Two other organisms which predispose to foot rot are the *Fusiformis necrophorus* and the *Spirochaeta penortha*.

In untreated or wrongly treated cases secondary bacteria — *Corynebacterium pyogenes*, *Steptococci*, *Staphylococci* and even *Bacterium coli*—move in with severe and often disastrous results *(photo 2)*, leading to invasion by maggots *(photo 3)*.

Mud or wet, badly drained pasture predispose to a disease flare up, but the mud itself will not produce foot rot unless the germ is there, and it can only live on the pasture or in the mud for an absolute maximum of 14 days.

It is acutely contagious. If the disease is not controlled, up to 100 per cent of the flock can become infected.

1

Symptoms

The picture usually is one or two lame sheep followed in a few days by several more.

What usually happens in an affected

2

3

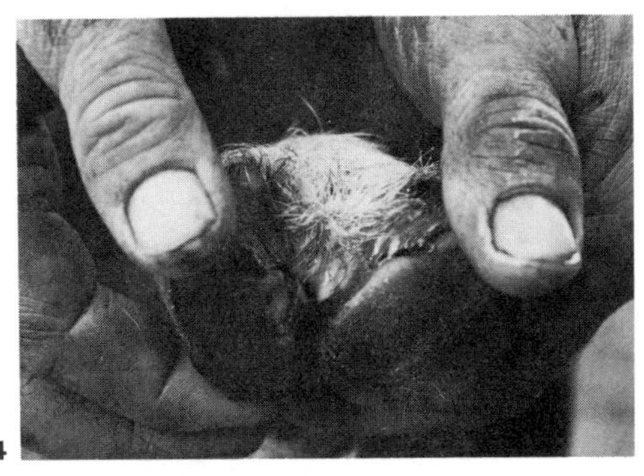
4

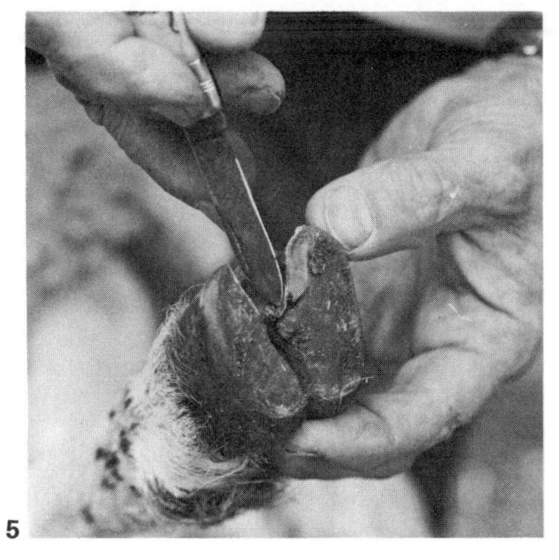
5

foot is that a moist mild inflammation develops between the claws *(photo 4)*. This early condition is sometimes called 'scald' and is caused either by the *Fusiformis necrophorus* or the *Spirochaeta penortha*. Very quickly a break appears between the skin and the hoof horn *(photo 5)*; the *Fusiformis nodosus* moves in and rapidly causes a separation of the horn from the underlying tissues. The Fusiformis does in fact cause death of the tissues and this leads to a characteristic stinking discharge.

Needless to say, untreated cases suffer considerable pain and lose weight rapidly, with the lameness becoming progressively worse *(photo 6)*.

Treatment

As soon as lameness is detected in a flock, the problem should be tackled as quickly as possible.

Bring the entire flock, not forgetting the ram, into a concrete yard. If you haven't got a concrete yard, put them in pens floored with a thick layer of straw bedding.

Lay out one or two decent foot knives, a pair of secateurs and a sharp scalpel (your veterinary surgeon will supply the scalpel with a packet of spare blades). You also need a bucket of disinfectant for dipping the cutting tools between each foot and a brush for cleaning the feet *(photo 7)*.

6

7

8

9

Taking one sheep at a time examine each foot using the brush, then the secateurs *(photo 8)*, then the knives *(photo 9)*.

Search carefully for infected animals. There are three types to look for:

1. The obvious type with lameness and under-run horn covering up the infection *(photo 10)*.
2. The mis-shapen hoof in which there is a small pocket of infection under the horn, usually at or near the point of distortion *(photo 11)*. This is the common type of 'carrier' sheep.
3. Cases of 'scald' where the hoof is normal but the skin between the claws is hairless, mildly inflamed and moist *(photo 12)*.

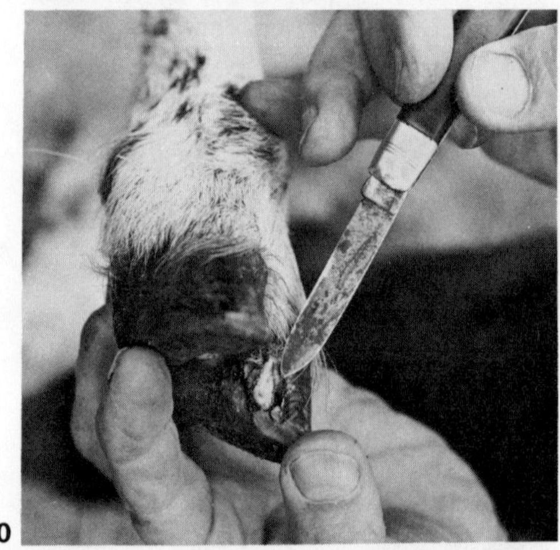

10

After each sheep has been pared and treated, the foot parings should be swept up and the instruments rinsed in the antiseptic solution. This is often forgotten.

11

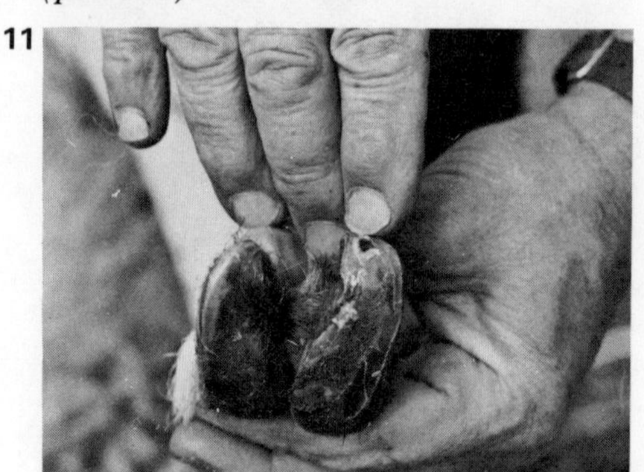

12

13

Separate all sheep in any of these three categories from the healthy stock. Now drive the clean sheep through a 10 per cent formalin or 10 per cent copper sulphate foot-bath, keeping the feet immersed for at least 1 minute. Afterwards leave them on the dry concrete yard for an hour or two *(photo 13)*; then turn them away on to a clean pasture (*i.e.* a pasture which has not carried sheep for at least 14 days).

Next, set to work on the affected animals. Pare away all the dead tissue and overlying horn, paying particular attention to the front of the foot. It is absolutely vital to expose all the areas of infection so that the antiseptic or antibiotic dressing can destroy the germ.

Start with the secateurs, then the knife, but, as soon as the infection is exposed, use the scalpel; always cutting from *within outwards (photo 14)*. This simple hint will avoid bleeding and allow a much cleaner and more thorough dissection.

When all the under-run tissue is exposed, dress with one of the modern antibiotic or antiseptic dressings *(photo 15)*. Most modern antiseptics will destroy the foot rot germ, but (and this is important) the old-fashioned caustic preparations such as 'butter of antimony' are inefficient and painful. They retard healing and should never be used. Ten per cent formalin is cheap and effective.

If, as sometimes happens, the foot is invaded by maggots a special anti-fly dressing, like sheep dip, should be combined with the antiseptic or antibiotic *(photo 16)*.

If the foot is extensively damaged, wrap it up in cotton wool and bandage *(photo 17)*.

14

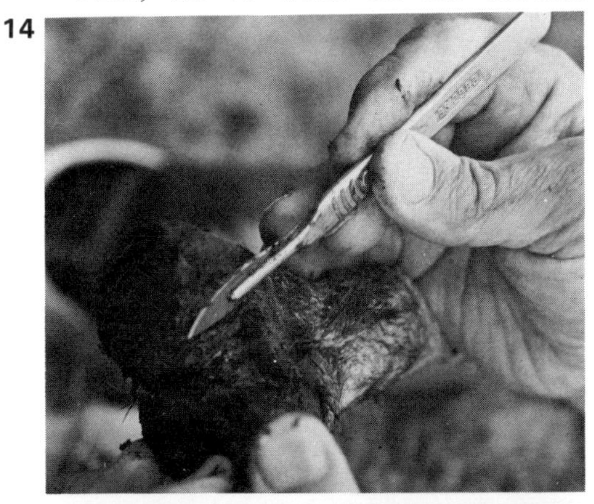

15

16

17

When all the infected sheep have been treated, turn them on to a separate clean pasture well away from the unaffected sheep.

Examine and treat the affected animals again in 2 or 3 days, and thereafter once a week until they are cured. Mostly two dressings will be sufficient.

Any new sheep or any which leave the farm and return should be examined, pared and segregated for at least 14 days. At the end of that time, if the feet are still healthy, they should be immersed in the formalin bath before the sheep join the main flock.

18

Prevention

There is a new highly efficient vaccine against foot rot *(photo 18)*. Two doses are given initially at an interval of 1 month and thereafter the immunity is boosted by a dose each year.

The vaccine is excellent but not infallible and will only prove 100 per cent effective when combined with the first-class routine husbandry described in the 'Treatment' paragraph above.

The vaccine can be used as a treatment, but obviously its main role in the future will be as a preventative, and as such it should be used by every conscientious sheep farmer as a vital part of commonsense routine husbandry.

29
Additional Foot Conditions

(a) Foot and Mouth Disease

DURING a foot and mouth outbreak, early diagnosis of the disease is usually very difficult in sheep. This is because the only obvious clinical sign may be lameness.

Cause
A filtrable virus which multiplies and spreads rapidly.

What Happens
During an outbreak the sheep eat the virus *(photo 1)*. It gets into the bloodstream and travels to the mouth, nostrils, feet and occasionally the udder. On these sites the virus multiplies underneath the skin to produce blisters which burst and discharge millions of the viruses on to the pasture.

1

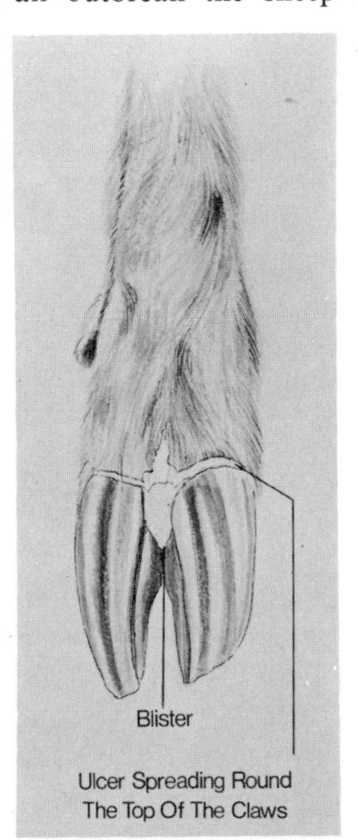

The blisters on the udder, nostrils and mouth are usually small and difficult to see. In the mouth they form on the dental pad, inside the lips or on the tongue.

Symptoms
The flock becomes generally dull and many of them stop grazing. Shortly afterwards widespread lameness develops. The

lameness is due to the tenderness of the ruptured blisters and to secondary ulceration and infection *(see sketch)*. Close examination will reveal the blisters or ulcers around the top of and between the claws or at the base of the supernumerary digits.

There may be similar lesions on the dental pad, inside the lips, or on the tongue; with a very occasional blister on the udder.

Treatment
Any suspect case must be reported to the police or to your veterinary surgeon immediately. Having done so, you must on no account leave your farm until the authorities allow you to do so.

Control
If a case is confirmed, the entire flock and all contact cloven hoofed animals will be slaughtered.

(b) Redfoot

This is a condition apparently limited to the Scottish Blackface breed. It affects the newly born lambs.

Cause
The cause is unknown, though some scientists suspect a congenital factor.

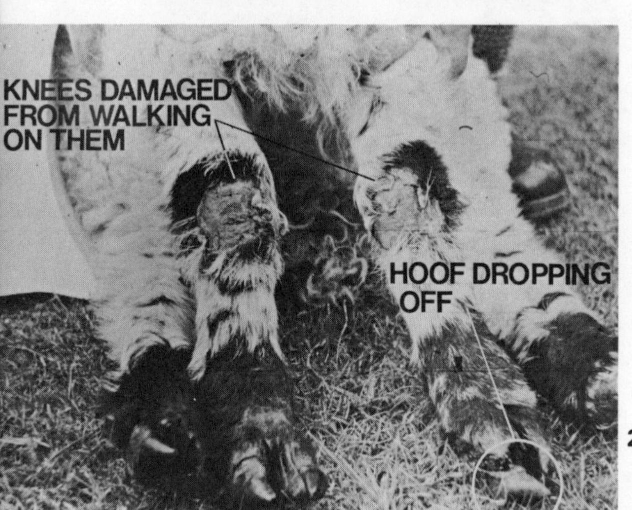

2

Symptoms
The horn of the hooves becomes loose or drops off *(photo 2)* exposing the red sensitive laminae underneath—hence the name Redfoot.

The lamb is in considerable pain. It walks on its knees, is unable to suck and rapidly dies of starvation.

Treatment
The best treatment is immediate painless euthanasia (destruction).

(c) Strawberry Foot Rot

This is a contagious disease affecting sheep of all ages but especially lambs.

Cause
A fungus belonging to the Dermatophilus species.

Symptoms
Outbreaks usually occur when there is a wet spell during the summer.

The skin between the top of the foot and the knee or the hock is inflamed and moist. It becomes thickened and scabs over. When the scabs are removed or drop off red shallow haemorrhagic ulcers are left—hence the name Strawberry Foot Rot.

Treatment
A 3-day course of injections of mixed penicillin and streptomycin, combined with the local application of a mild caustic solution such as 0·5 per cent zinc sulphate, will clear up most cases.

If large numbers are affected, they should be passed through a deep foot bath, containing 0·5 per cent zinc sulphate solution, once a week for 3 weeks.

Prevention
The routine husbandry advised for the control of foot rot should be sufficient to keep the fungus under control.

(d) Foot Abscess

It is my experience that most foot abscesses in sheep are secondary to neglected foot rot *(photo 3)*. However, they can and do occur independently of foot rot and it is therefore important to recognise them and treat them correctly.

Cause

The germs involved are usually the *Fusiformis necrophorus* and the *Corynebacterium pyogenes.*

Symptoms

Severe lameness. The area above the foot is usually hot, swollen and painful and the digits are spread.

3

Treatment

A 5-day course of penicillin injections (6 cc per day of 300,000 units per cc) combined with local treatment such as daily soaking in hot water containing Epsom salts (one tablespoonful to the pint), or kaolin poulticing.

As soon as the abscess bursts the pain will subside, but if the course of penicillin is not given the abscess is likely to recur.

Prevention

Control foot rot by vaccination and routine good husbandry, and foot abscesses will be a rarity in the flock.

INFERTILITY AND ABORTION

30
Infertility

FLUSHING THE ewes immediately prior to mating goes a long way to prevent infertility *(photo 1)*.

In Great Britain and Ireland there is no known disease that produces infertility in ewes. Where a number of ewes run barren on lowland farms, the fault usually lies in the management or with the tup.

The heavy feeding of ewes immediately after mating is thought to be a possible management cause.

When the tup is at fault, he is usually being asked to do too much work.

On hill sheep farms there is a known relatively high incidence of barren ewes; here the cause is generally accepted to be nutritional—certainly no disease bacteria have been implicated.

In Australia and New Zealand ovine brucellosis produces infertility in rams.

1

31
Abortion

ABORTION IN ewes can be a serious problem *(photo 1)*. There are a number of different types of abortion, and every flockmaster should be aware of these and especially informed on how the various bacteria get into the flock and what damage they are likely to do. In addition, he should know exactly what precautions he can take for subsequent lambings.

First of all, there are the specific abortions, *i.e.* those caused by specific bacteria or organisms. For easy reference I tabulate what I consider to be the essential information.

Specific Abortion	*Cause*	*How it Gets into a Flock, and Other Important Facts*	*What the Shepherd Should Do*
Enzootic Abortion ('Kebbing abortion')	A large virus called a Bedzonian agent	Purchase of infected sheep of any age, including lambs. Within the flock the disease is spread only at lambing time *but infection is not usually spread by the ram.* Enzootic abortion remains in a flock once it is introduced and abortion rate will vary from 5 to 25 per cent of all the ewes; abortions occurring 2 to 3 weeks before full term	There is a first-class vaccine available. All breeding ewes should be vaccinated immediately—twice the first year and once a year thereafter *(photo 2)*

1

2

31–ABORTION *(continued)*

Specific Abortion	*Cause*	*How it Gets into a Flock, and Other Important Facts*	*What the Shepherd Should Do*
Vibrionic Abortion	A spirochoet—the *Vibrio foetus intestinalis*	Purchase of infected carrier sheep, or it can be introduced by wild fauna. The infection is taken in by the mouth and is *not* spread by the ram. When introduced, the disease can cause up to 60 per cent of the ewes to abort, though in most of the infections I have had to deal with the losses have been between 10 and 20 per cent, abortions occurring about a month before term	Keep the flock self-contained. One attack apparently gives a long immunity and doesn't appear to affect the fertility of the ewes for the following season. However, 'carriers' will be created and any bought in sheep will probably become infected and are liable to abort their first crop
Salmonella Abortion	A bacterium—the *Salmonella abortus ovis*	Purchase of infected sheep of any age. Occasionally the bug is carried by vermin. As with Vibrio, the germ has to be eaten and it is not spread by the ram. First year losses average about 10 per cent though severe attacks can bring the losses to 40 or 50 per cent, abortions occurring up to 6 weeks before term	Again, keep the flock self-contained. One attack confers a powerful immunity and the majority of ewes will breed successfully the following year. Any purchased ewes would be exposed to infection by carriers
Toxoplasmosis	A toxoplasma—the *Toxoplasma gondii*	The organism is wide-spread in nature and can be introduced by a number of 'carrier' animals. Man can also carry and spread it. Losses rarely rise above 10 per cent. Abortions occur at any time during the second half of pregnancy	Keep the flock self-contained. The tendency is for an immunity to be established. Any new sheep introduced will be exposed to infection from carriers
Q Fever	A virus	Again, the causal agent can be spread by many carrier animals, vermin and man. Abortion losses up to 10 to 15 per cent	Keep the flock self-contained. A powerful immunity develops, but any incoming sheep will be susceptible to infection

3

Secondly, there are the non-specific abortions—those not due to microbes. Strangely enough, the total numbers of these far exceed the losses from specific abortion.

The causes are complex and difficult to pinpoint, but probably the most important factor in keeping their numbers to a minimum is the correct feeding of the in-lamb ewes *(photo 3)* during the last 5 or 6 weeks of pregnancy (see Pregnancy Toxaemia, p. 15).

Finally, there are the indirect abortions due to certain conditions in the mother that lead to death of the lambs in the uterus. In this country such deaths occur in louping ill, congenital goitre, tick-borne fever, and pregnancy toxaemia. In other countries foot and mouth disease, Rift Valley fever, blue tongue and ovine brucellosis can all cause death of the lambs inside the ewes.

Obviously, it is wise in all cases of abortion to consult a veterinary surgeon. He, with the help of a laboratory, will establish the correct diagnosis and will advise accordingly.

OBSTETRICS

32

How Long Should Ewes be Kept?

SOME TIME ago I asked a farming friend of mine—one of the most successful sheep farmers I know—how long he kept his ewes. He replied, 'Until they have been dead at least three days.' This laconic retort, to my mind, sums up the commonsense approach to successful sheep farming.

Provided the udder is sound and there has been no prolapse problems at the previous lambing, it is wise to keep the ewes as long as possible and that can mean many years. The ewe pictured here was 27 years old at the time of the photograph. She had a crop of lambs every year for 24 years and her breeding career only came to an end when she required a caesarian section.

The advantages of keeping these old ewes are obvious. Apart from the low capital outlay, the ewes acquire a powerful resistance against all the diseases encountered on that farm. For the same reason one should always breed one's own replacements.

Bought-in sheep have little or no resistance to the resident local bugs and will take several years to settle down. This fact has been illustrated again and again during restocking after foot and mouth outbreaks.

33

A Guide to Good Lambing

IT IS easier to kill a ewe by lambing her prematurely or roughly than by shooting her with a humane killer. If you want to keep lambing losses to a minimum, the first basic fact to remember is that it always pays to give Nature a chance *(photo 1)*.

The percentage of malpresentations in the ewe is comparatively small. In any case it is quite safe to leave a ewe in labour for at least 4 hours before examination. If left alone, then the vast majority of cases will lamb by themselves with no trouble whatsoever.

Even when the lamb is straight, it is wrong to pull it away quickly because the sudden temperature change and atmospheric pressure change will often kill it. Every experienced shepherd has seen this happen: an apparently normal lamb is pulled away quickly—it takes one gasp and dies. Such death is due to shock.

This is a simple but nonetheless tremendously important point well worth bearing in mind. *When the lamb is straight, never interfere unless the ewe ceases to make progress.*

2

1

Apart from patience there are four golden rules at lambing time.

1. Cleanliness.
2. Getting the ewe into the correct position.
3. Lubrication.
4. Careful manipulation.

Cleanliness

I know cleanliness in the field is not always easy, but I believe that, wherever possible, no examination should be attempted in the field. The suspect ewe should be loaded up and taken to the farm building *(photo 2)*.

Transport is so easy on most farms that it's ridiculous to attempt interference

3

4

5

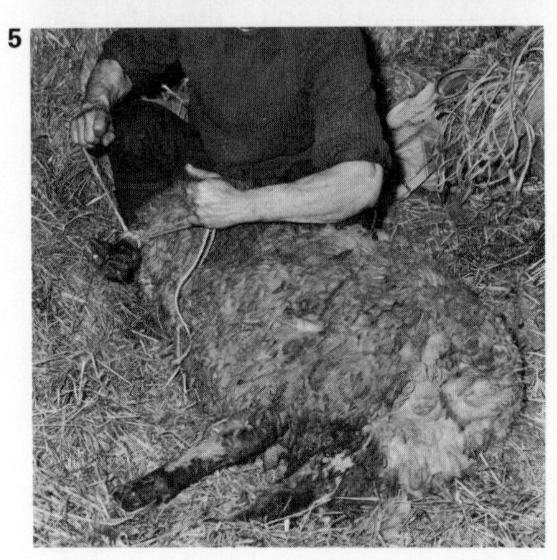

6

outside. At the building, the patient should be taken out, and prepared for the operation on a clean bed of straw. And, of course, it is preferable and much more comfortable if the examination can be done under a roof.

All the wool and dirt should be clipped from around the vulva *(photo 3)*. This is a very useful hint, because without clipping, real cleanliness is impossible. The entire area should be thoroughly scrubbed with warm water, soap, and non-irritant antiseptic *(photo 4)*.

Getting The Ewe Into The Correct Position

Lay the ewe on her side and tie the forelegs together above the fetlocks with thick bandage or string *(photo 5)*. Roll the ewe on to her back and get an assistant to stand astride her facing the tail *(photo 6)*.

Holding each hind leg above the hock, the assistant should lift the hind-quarters as high as possible, and a bale or filled sack can then be put against the ewe's spine to take most of her weight *(photo 7)*. In addition, a $\frac{1}{4}$-inch rope can be fixed to

7

8

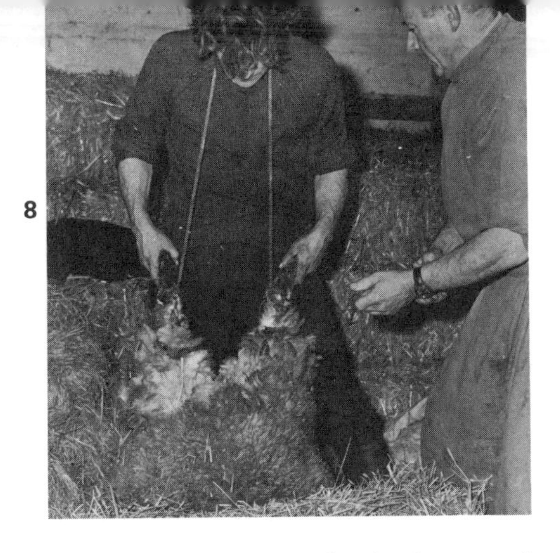

one hind leg above the hock, passed around the back of the assistant's neck and shoulders, and tied above the other hock *(photo 8)*.

I know that the hill man rarely has the luxury of an assistant, nor is it possible for him to take his patient back to a building. But in his case, an effective improvisation is easy, using the $\frac{1}{4}$-inch rope—with the forelegs tied, of course. This time the rope is passed over a fence post and fixed above both hocks *(photo 9)*.

I think it is always worth while trying to get the ewe into this position because with the ewe like this, the job of examining for presentation, and of correcting malpresentation, is infinitely easier.

Lubrication

The shepherd should now again wash the vulva region, and insert copious quantities of soap flakes and warm water into the vagina *(photo 10)*. Some people use liquid

9

10

11

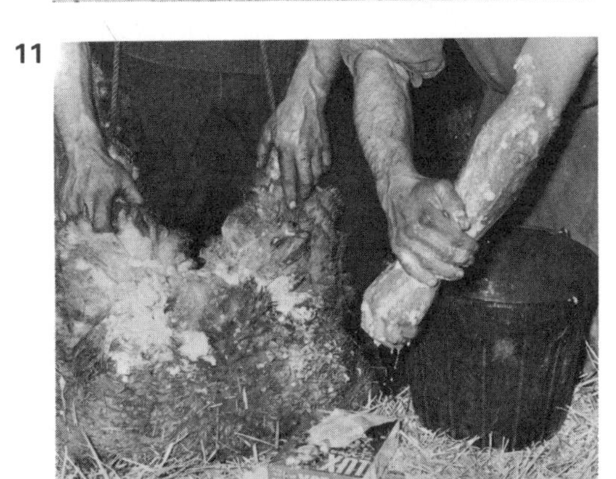

paraffin or linseed oil; for me soap flakes comprise the ideal lubricant—and I always fill the vagina with them.

That done he should scrub his hand and arm, and if possible, lubricate them with an antibiotic cream and still more soap flakes *(photo 11)*.

The hill shepherd, with no access to farm buildings or warm water, has to rely entirely on his lubricant cream. First of all, he should fill the vagina with it and smear it over the surrounding area; then generously coat the hand and arm *(photo 12)*. One important point: the old-

12

13

14

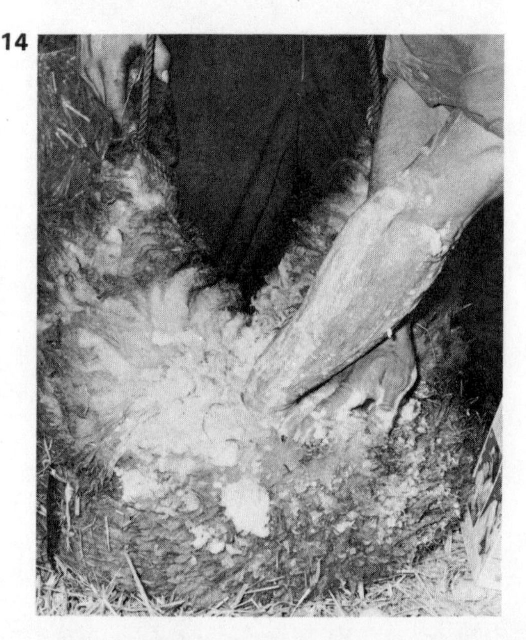

fashioned lambing oils should be avoided because, in the main, they are irritant and do more harm than good. Every hill sheep farmer should provide himself with adequate supplies of non-irritant lubricant cream. It will be money well spent.

Careful Manipulation

The hand should be introduced patiently and very slowly, taking great care not to lacerate the passage *(photo 13)*. I am certain that every really experienced shepherd will agree wholeheartedly with me when I say that the ewe's inside is so delicate that one just can't be too gentle.

All the work done inside should be done with the same care, constantly renewing the lubrication when necessary. It's surprising how easily seemingly hopeless tangles can be sorted out in this position provided the shepherd is patient and gentle. *I repeat, he can't be too gentle.*

As an indication of the amount of pressure he should use, I think he should move his hand about as though inserting it between the folds of delicate silk *(photo 14)*.

If the neck of the womb is only partially opened, the shepherd may well be dealing with the condition known as ringwomb (see chapter on Ringwomb), and he should never attempt to pull the lamb through the incompletely opened cervix. If possible, he should pass the case on to a veterinary surgeon immediately. If the presentation is correct, *i.e.* the two forefeet and the head of the lamb presented and the cervix fully dilated, then the shepherd should leave the ewe alone for at least another 4 hours. After that time, if the ewe still has not lambed, he should send for his veterinary surgeon.

34

Ringwomb in Ewes

RINGWOMB OCCURS when the cervix (the entrance to the uterus) does not open properly during labour *(photo 1)*. There is a true ringwomb and a false ringwomb.

In a true ringwomb, relaxation of the cervix proceeds so far and no further, despite the application of all known treatments.

In so-called false or partial ringwomb, relaxation occurs with one of the recommended procedures.

Causes

Every shepherd knows about ringwomb, but not all of them understand exactly what it is or what can or should be done when they come across it. We veterinary surgeons know what it is and what to do—but we still don't fully understand the condition because the precise cause has not yet been established. Most of us think that the majority of cases are due to some hormonal deficiency, and much of the research work being done is directed towards proving this supposition.

1

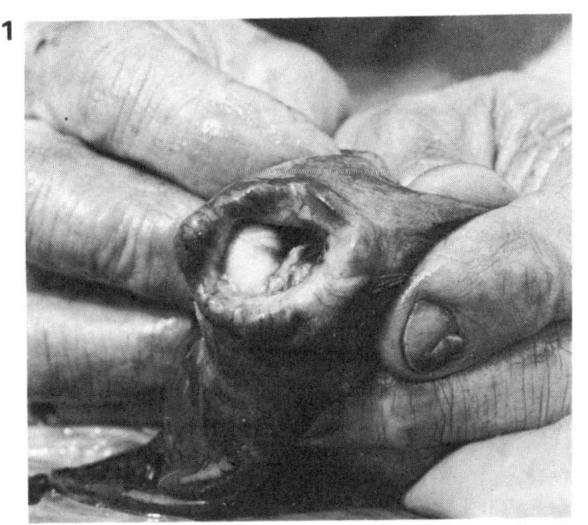

2

Symptoms

When the ewe is examined, after, say, 4 hours of ineffectual straining, only one or two fingers can be passed through the cervix into the womb *(photo 2)*, even though the two forefeet and head of the lamb may be presented quite normally. The outside edge of the cervix usually feels hard and unyielding, almost like an

3

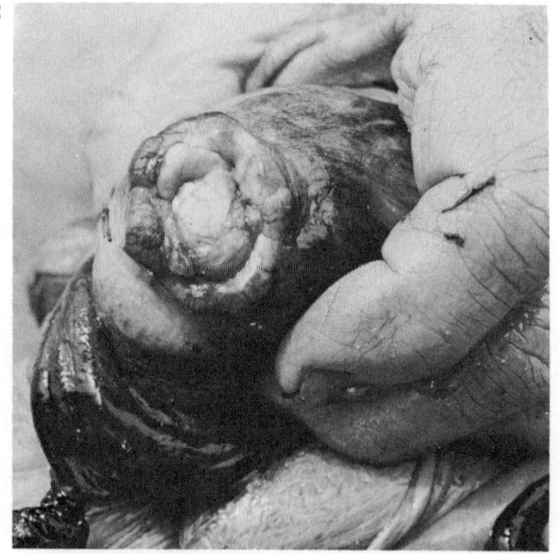

4

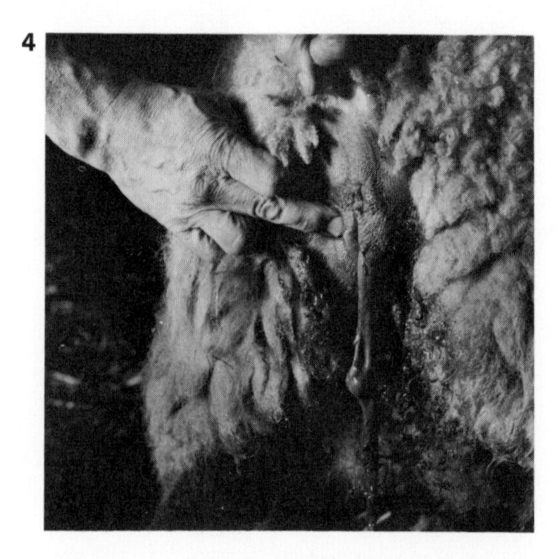

extended rubber ring—hence the name 'ringwomb'.

The cervix itself is made up of a group of muscles pouched together round the uterine entrance *(photo 3)*. In a normal labour these cervical muscles relax and open up the entrance each time pressure is exerted from within. This pressure occurs intermittently during labour and is caused by the contractions of the muscles of the uterine wall.

Pressure on the inside of the cervix is exerted first of all by the so-called water bladder which surrounds the lamb, and thereafter, when the bladder bursts, by the pressure of the lamb's legs and head —mostly by the head—which exerts its pressure on the muscles at the top part of the cervix.

Ringwomb is not usually diagnosed until the ewe is examined, though I have found that a fairly consistent symptom is the appearance of dangling afterbirth with no sign of a lamb *(photo 4)*. Such cases will probably have been in labour for 4 or 5 hours, or even longer.

In order to examine the ewe correctly, she should be placed in the position for lambing (see Lambing) and the same precautions of cleanliness, lubrication and careful manipulation should be observed.

Is There Any Danger To The Lambs?

In all cases of true or false ringwomb, there is some danger to the lambs, and this increases the more the labour is prolonged. The continued ineffectual uterine contractions may cause an actual separation of the placenta, or afterbirth, from the cotyledons which form the afterbirth's attachments inside the womb. When this happens, the life-giving blood supply from the mother is cut off and dead lambs are inevitable.

Treatment

If the hard, unyielding band or 'ring' is there, then the best thing to do is to turn the case over to your veterinary surgeon. I have found in such cases that the most convenient idea for both parties is for the shepherd to run the ewe down to the veterinary surgeon's hospital or premises. Even on many hill farms this idea is practicable.

If, however, as happens in some of the more remote hill farms, veterinary surgeons are not easily accessible, then there are certain things that the shepherd should try.

First of all, using plenty of soap flakes *(photo 5)* and warm water he should pass one or two fingers through the ring and

5

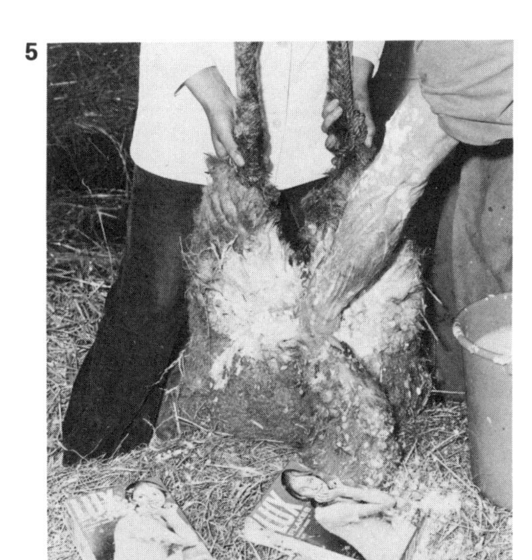

6

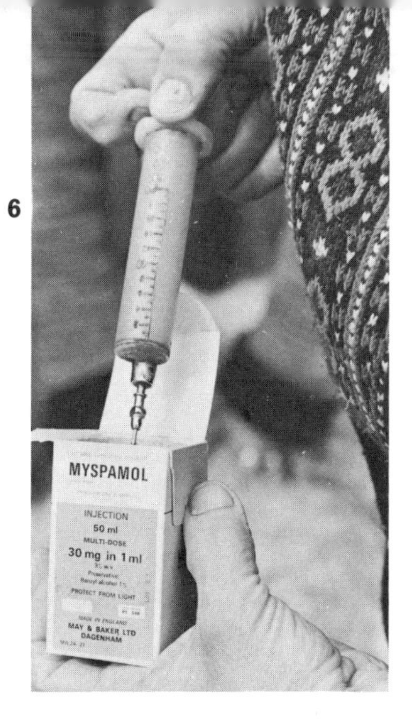

exert gentle pressure on the inside—particularly at the top—repeatedly filling the vagina and cervix with the flakes and warm water and working away quietly and gently for up to half an hour or even longer if progress is being made.

It is surprising how many 'ringed' cervices will relax and open up under this simple internal gentle digital pressure. Personally I like to keep my fingers moving round the inner ring and stopping at the top each time the ewe strains.

If *no* progress is made during the first 15 minutes of digital pressure, then an antibiotic should be injected and the ewe should be left for a further 12 hours. This, of course, presumes that the shepherd will be armed with antibiotic and syringe. In my opinion, all outland shepherds should be supplied with these by their veterinary surgeons for use in emergency. The antibiotic will help to prevent any dead lambs from putrefying.

If the sheep farm is particularly inaccessible, then the veterinary surgeon may supply a shepherd with his own more specific treatments, e.g. a muscle relaxant (May & Baker's MYSPAMOL is a typical example *(photo 6)*). This is injected intramuscularly and often acts in a spectacular manner within ½ to 1 hour *(photo 7)*. In somc areas certain hormone injections appear to be specific, and here again the veterinary surgeon will advise and liaise.

If patient digital pressure, time and Nature, and muscle relaxants and hormones all fail, then only two alternatives remain—emergency slaughter or Caesarean section.

Caesarean section is very successful, and most veterinary surgeons will do everything possible to keep the cost of the operation in line with the potential value of the ewe.

7

35
Caesarian Section

THIS OPERATION is often necessary in true ringwomb, though it may have to be resorted to when the foetus is oversize or abnormal or when there is an irreducible twist in the uterus. Needless to say, it should only be performed by a qualified veterinary surgeon, and I describe it here mostly for the benefit of veterinary students.

1

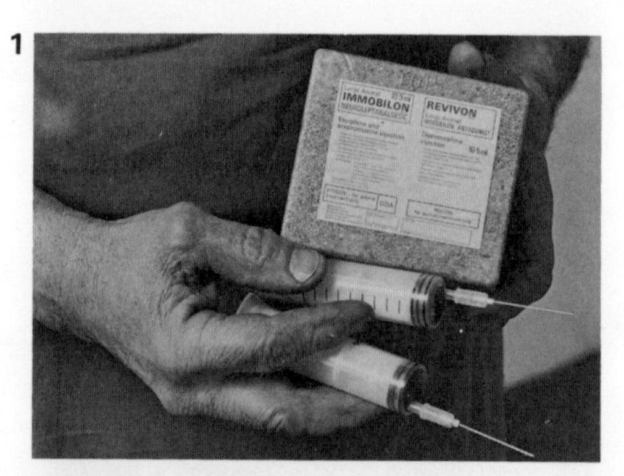

2

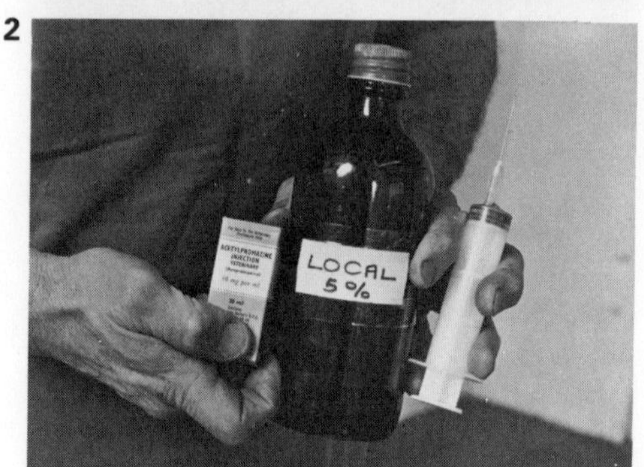

The Anaesthetic
Several modern general anaesthetics can be used, such as FLUOTHANE (ICI) or IMMOBILON and REVIVON (Reckitts) *(photo 1)*. Personally I prefer a local anaesthetic combined with a tranquilliser, since in my experience this technique offers the best prospects of obtaining live lambs *(photo 2)*.

The Site
The lower part of the left flank equidistant between the last rib and the point of the hip.

The Technique
The tranquilliser is administered intravenously, and the ewe is laid on a bench on her right side. The two fore and two hind legs are tied together *(photo 3)*.

3

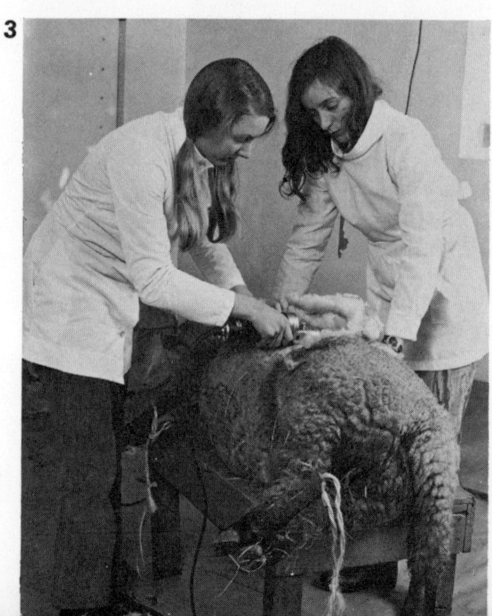

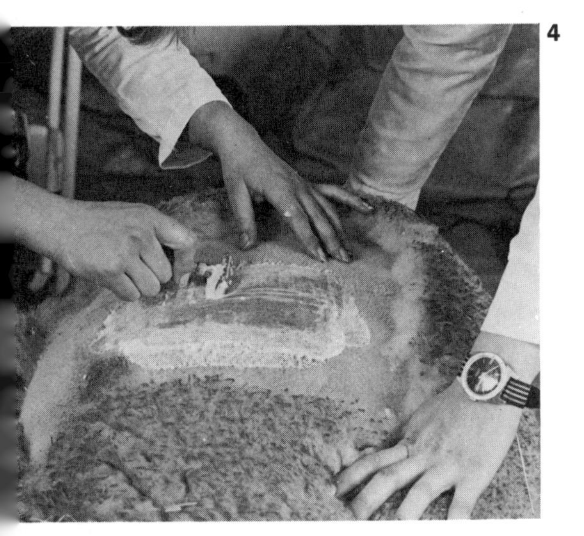

4

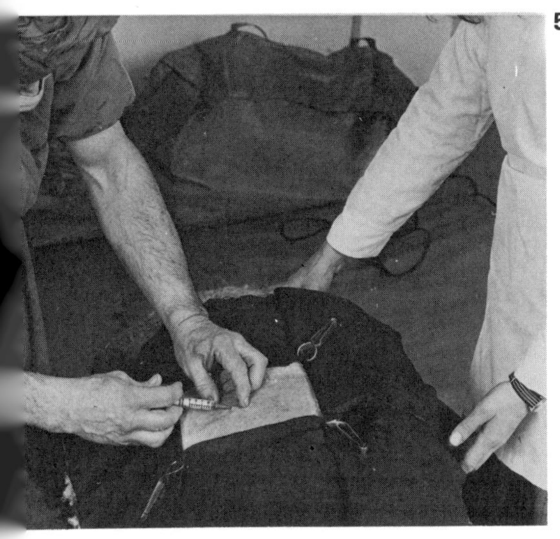

5

The entire left flank area is clipped, shaved *(photo 4)* and thoroughly scrubbed. It is then painted with a powerful skin antiseptic and marked off by sterile operation cloths.

Approximately 20 cc of a 5 per cent solution of local anaesthetic is infiltrated under the skin and into the muscles and peritoneum along the line of the proposed incision *(photo 5)*. At least 5 minutes should elapse before commencing the operation. This will allow the local anaesthetic to take full effect.

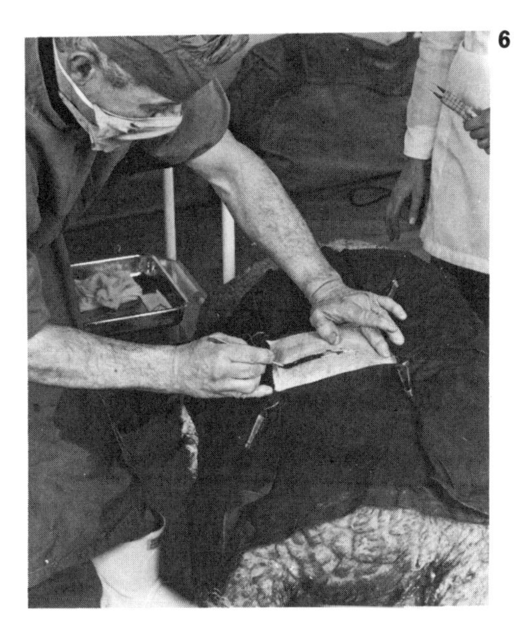

6

Having thoroughly scrubbed up, an incision, approximately 5 or 6 inches long, is made in line with the last rib and towards the lower flank *(photo 6)*. The incision passes successively through the skin, the superficial abdominal muscle, the deep abdominal muscle and the peritoneum *(photo 7)*. I find it preferable to use straight-bladed scissors, or a director, to cut through the peritoneum because of the danger of laceration of the viscera.

Now both hands and forearms are lubricated with an antibiotic oil or cream —I use intramammary antibiotics for this job.

7

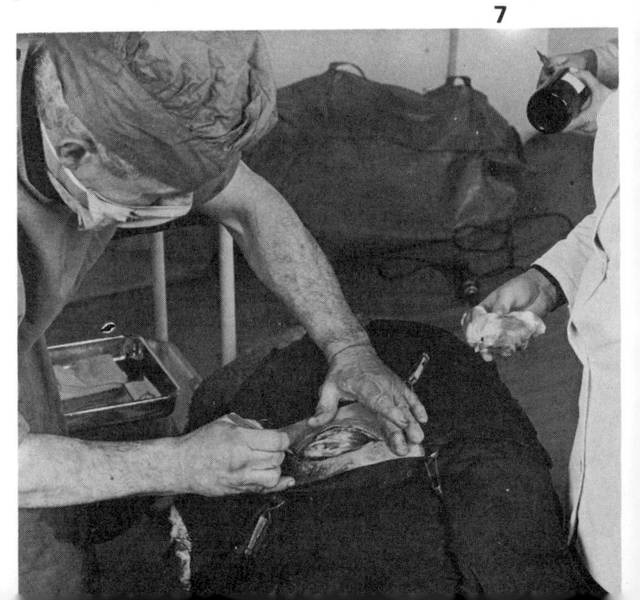

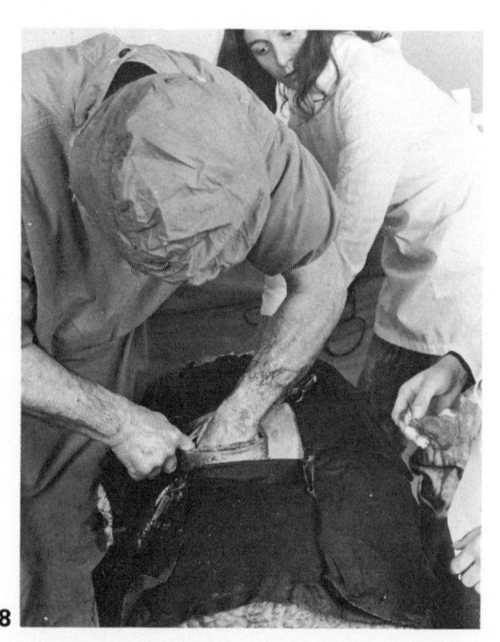

8

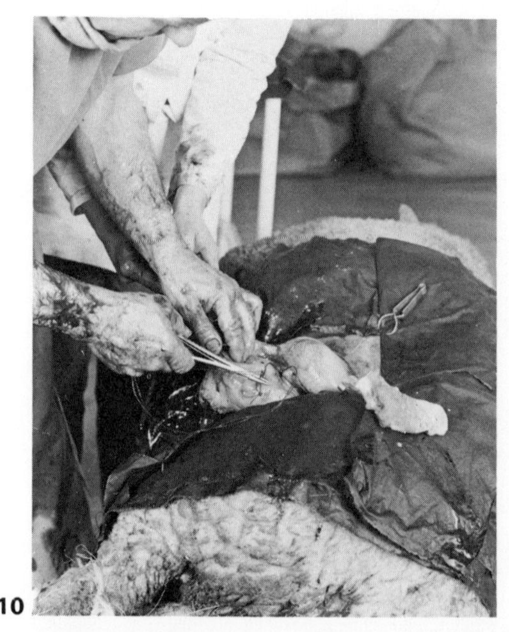

10

One hand is inserted into the abdomen and underneath the nearest pregnant uterine horn *(photo 8)*. Taking great care not to tear the uterus, the horn is lifted up gently until it is hard against the open wound.

With the free hand a controlled incision about 4 inches long is made in the uterine wall.

Still using the free hand, and with the other still pressing the horn upwards, the lamb's hind or fore feet are withdrawn from the uterus and the lamb is lifted carefully out *(photo 9)*. This lifting will bring the uterine wound outside the external incision and will leave the other hand free to assist in the withdrawal.

The remaining lamb or lambs are removed in exactly the same way, and the lips of the uterine wound are secured with tissue forceps. If possible, the afterbirth should be removed.

The uterine wound is then closed by a single continuous Czerny-lembert suture using double No. 3 or No. 4 catgut. I prefer this double thick catgut because it lessens the danger of uterine tearing *(photo 10)*.

The peritoneum and internal abdominal muscle are now closed by a continuous

9

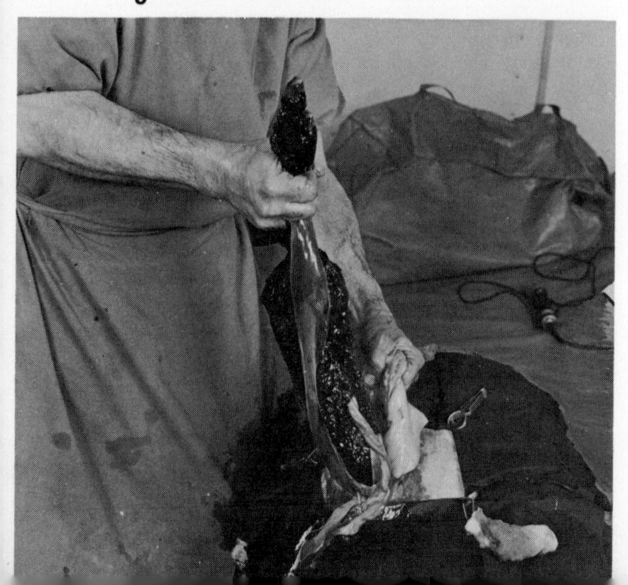

11

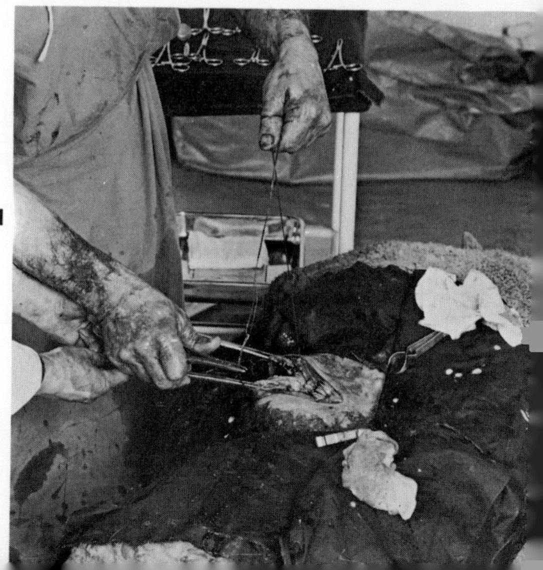

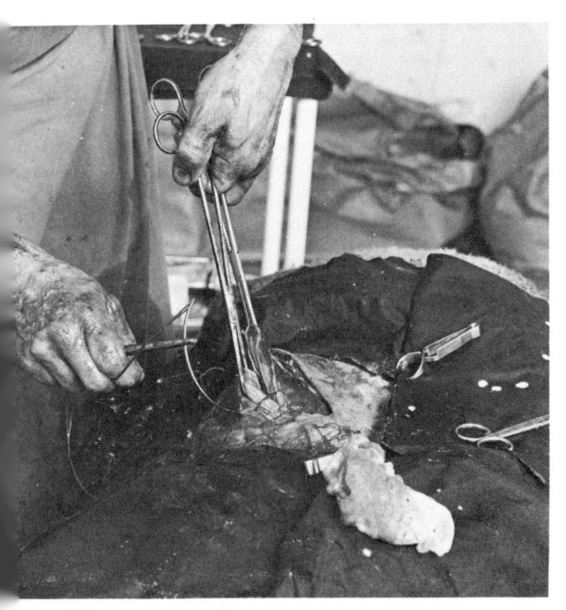

12

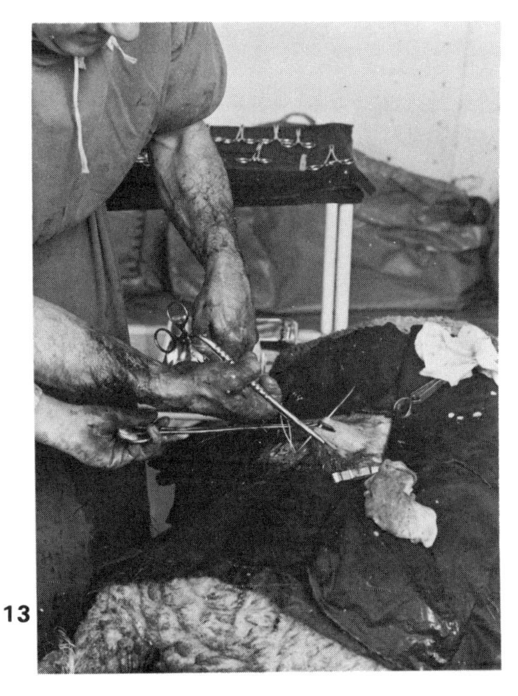

13

suture again using double No. 3 or No. 4 catgut *(photo 11)*. Usually I take the precaution of introducing some intramammary antibiotic into the abdominal cavity before closing the wound.

The external abdominal muscle is brought together, again by a continuous catgut suture *(photo 12)* and the skin wound is closed by two or three nylon or silk mattress sutures *(photo 13)*.

A final dusting of the wound with a sulphanilamide powder and an intramuscular injection of 6 cc (300,000 units per cc) of a long-acting antibiotic completes the operation *(photo 14)*.

Sutures are removed in 2 or 3 weeks.

Prognosis

Excellent, especially if the lambs are alive. Even when the lambs are dead the prognosis is good, provided ample antibiotic is used in the abdomen and wound, and an intramuscular antibiotic cover is maintained for at least 10 days.

14

36

Prolapse of the Cervix

THERE ARE two types of prolapse—the prolapsed cervix (or neck of the uterus) and the prolapsed uterus (the entire womb). The former occurs before lambing, and the latter after. The prolapsed cervix, which is by far the more common, incorporates the ewe's vagina and occasionally the bladder also *(photo 1)*.

Cause

Prolapsed cervix appears to be one of the penalties of top-class farming.

In the non-pregnant ewe the rumen or first stomach alone may occupy up to three-quarters of the entire abdominal cavity, while the other three stomachs do more than their share towards filling up the remainder.

This means that, under natural conditions, there is just about enough room left for the development of a good single lamb. Twin lambs stretch the capacity to the limit, and how triplets find accommodation is little short of miraculous. In other words, the cervical and vaginal prolapse is due primarily to a lack of room.

There are other contributory causes such as excess fat, stretching of the broad ligament of the bladder, and occasionally constipation; but all these only play their part when there is overloading of the abdominal and pelvic cavities. Without that overloading, they are of no significance.

The better the management at flushing and tupping time, the greater the percentage of twins and triplets. The better the feeding, the greater the growth of the lambs. Hence my certainty that cervical prolapse in the ewe is a penalty of good farming. It is a simple matter of trying to squeeze a quart into a pint pot.

Complications

When two average-sized lambs are incorporated in the prolapse, the vaginal wall usually stands the strain for a reasonable time—but if the lambs are well grown and vigorous, then rupture occurs and death ensues *(photo 2)*.

1

2

3

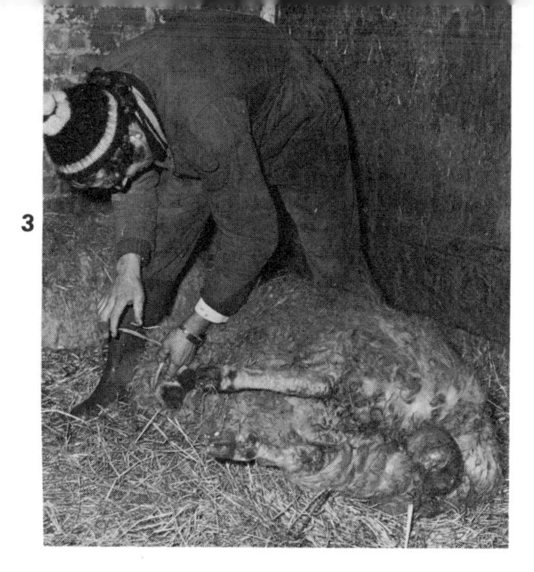

4

Treatment

The first essential is to get the ewe into the correct position for the return of the prolapse.

The technique is similar to that adopted for lambing, that is, the ewe should be turned on to her side and then rolled on to her back (the front legs can be tied together *(photo 3)*). An assistant should stand astride her facing the tail *(photo 4)* and, holding each hind leg above the hock, should lift the hind-quarters as high as possible. A bale of straw or hay or a filled sack can be placed against the ewe's back to take most of the weight *(photo 5)*.

5

6

If an assistant is not available, then the shepherd himself can pull the hind-quarters up by fixing one end of a $\frac{1}{2}$-inch rope above one hock and passing the free end around the back of his neck and shoulders to fix above the other hock. He can then lift the ewe and hold it comfortably by straightening his back. It is not a bad idea to do this in any case, even when help is available *(photo 6)*.

7

If the shepherd wants to work more freely, he can fix the rope around a post or to the top of a stone dyke.

Getting the ewe into the correct position is undoubtedly the most important part of the technique of prolapse return. The weight of the abdominal contents is transferred to the diaphram, the ewe strains with less power, and more room is available for an easy return.

8

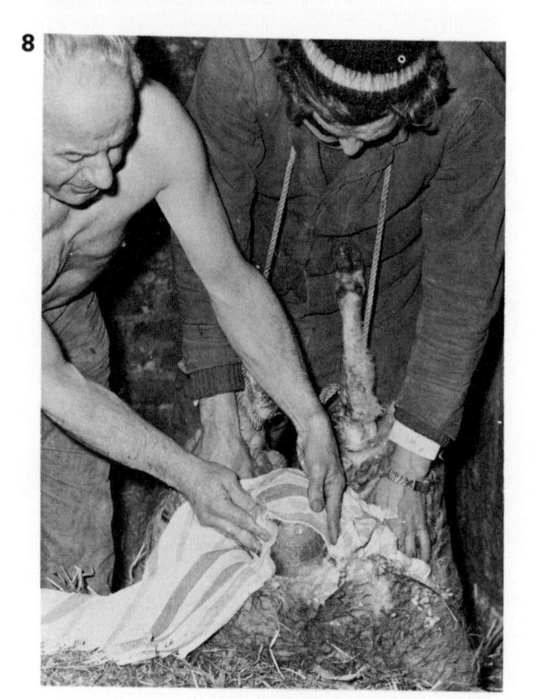

9

General cleanliness is very important. So use plenty of hot water, soap or, better still, soap flakes *(photo 7)* and some non-irritant antiseptic. After washing thoroughly, it is advisable to dry the prolapse and smear it with a non-irritant antiseptic or antibiotic cream before replacing it in the correct position. A rough towel is ideal for drying since it removes any remaining portions of dirt without damaging the uterus *(photo 8)*.

I sometimes use dry sulphanilamide powder as a final dressing because this penetrates into the surface of the prolapse and provides ideal protection against infection *(photo 9)*.

During the replacement gentle pressure should be applied constantly and increased between strains, holding the bulk or all of the prolapse in the palm of the hand, or in the palms of both hands, depending on the size of the prolapse *(photo 10)*.

10

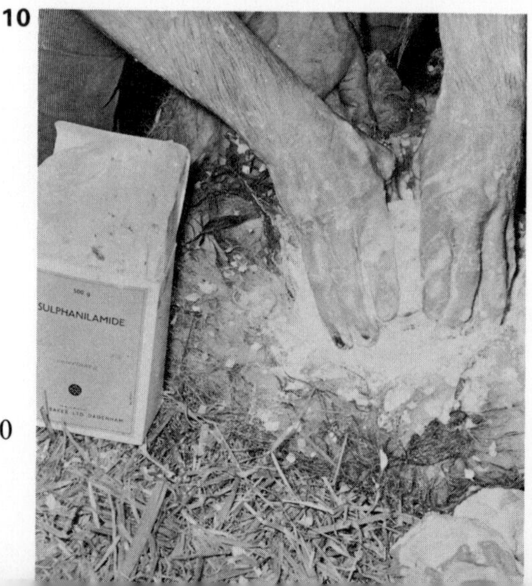

11

12

And now the method of stitching. I have found the tying of the wool, the insertion of wire retainers, and the ordinary type of stitching completely unsatisfactory.

There is one and only one infallible method of stitching. It comprises a single stitch, which we call a mattress or deep retention suture *(photo 11)*. A good stout needle and strong silk or nylon should be used. The method of inserting the suture is illustrated in the three sketches below.

The needle is passed as deeply as possible through both sides of the top of the vulva (A), is brought down to the bottom of the vulva (B) and again as deeply as possible (but avoiding the urethra or hole into the bladder) is passed back right through both sides. A reef knot is now tied as tightly as possible on the side of the original entry of the needle (C).

The suture should be left in position until the ewe lambs. It is not necessary to sit up and wait for this, but it is a good idea to keep a special watch so that the stitch can be removed as soon as the lamb's feet appear.

To remove it, take the free ends of the silk just above the reef knot, pull outwards and cut below the knot.

Can Cervical Prolapse Be Prevented?

I don't think it is possible to prevent cervical prolapse—certainly not without the risk of losses from twin lamb disease and metabolic disorders. Flushing, *ad lib* hay, and an ascending plane of nutrition are a 'must' in modern well-managed flocks.

It has been suggested (and in some parts of the country it is the practice) that hay should be restricted or withheld during the last month of pregnancy. The idea is to use as a substitute a smaller bulk feed of a high energy ration containing adequate fibre. The supporters of this theory justify it by saying that post-mortem examinations on pregnant ewes have shown that the rumen is so compressed as to be too small to accommodate hay. To this I would point out that post-mortem examinations can only be done on *dead* sheep, and the ones quoted probably died of starvation.

It is very wrong to deprive the pregnant ewe of ad lib *hay (photo 12)*. Nature will ration the intake according to available accommodation, but it is essential that sufficient natural fibre is provided to

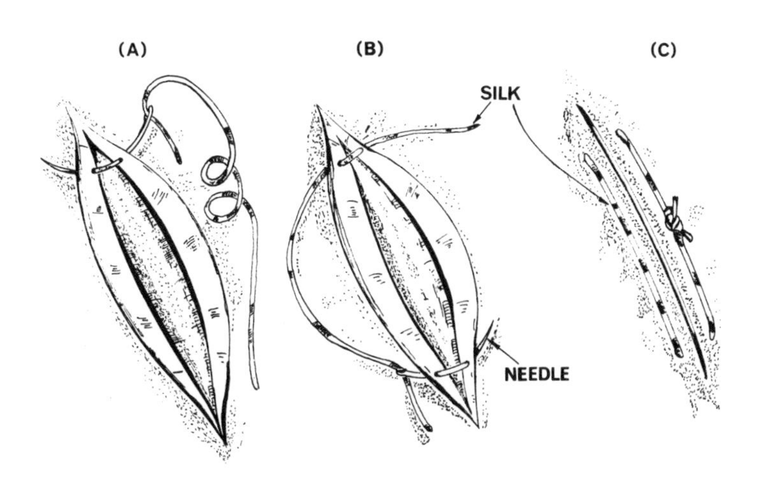

retain the concentrates long enough in the rumen, to allow the production of the fatty acids that go to keep the ewe warm and to provide energy for the growth of the lamb. Otherwise, the ewe will turn to her body reserves of fat and twin lamb disease will result.

Perhaps there is just one realistic practical step that might be tried—the avoidance, so far as possible, of a fresh new pasture during the last month of pregnancy. If there is any grass during this stage, the tendency is for the pregnant ewe to gorge.

In my opinion it is unwise to keep a ewe after she has prolapsed.

All cases should be carefully marked and culled before the next breeding season.

37

Prolapse of the Uterus

THIS OCCURS after lambing, usually within a few hours, though occasionally it can occur several days after. The entire womb, turned inside out, hangs from the vulva *(photo 1)*.

1

Cause

There are three causes:

1. *Calcium deficiency*

When the ewe is suffering from lambing sickness (or milk fever), the deficiency of calcium causes a loss of tone in the uterine and cervical muscles. At the same time, the patient is often constipated and the reflex straining to pass the droppings brings about the prolapse *(photo 2)*.

2. *Inversion of the tip of the pregnant horn*

Sometimes during labour the tip of the pregnant horn may become inverted *(photo 3)* or turned inwards (rather like when the toe of a sock is pulled in to the foot). When this happens, the ewe will continue to strain incessantly till the uterus prolapses.

2

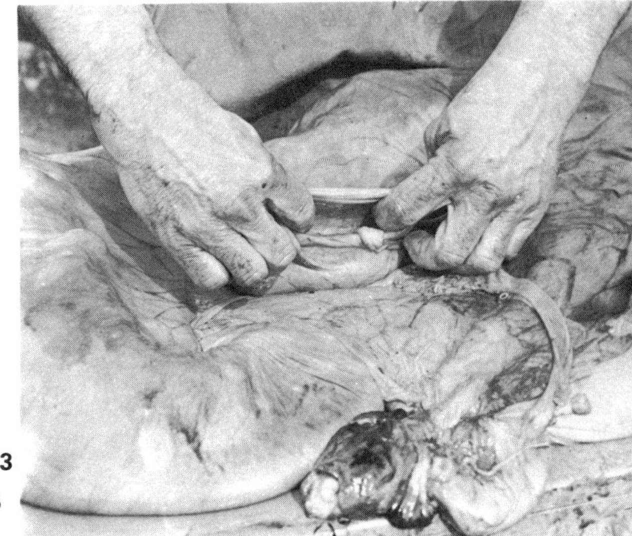

3

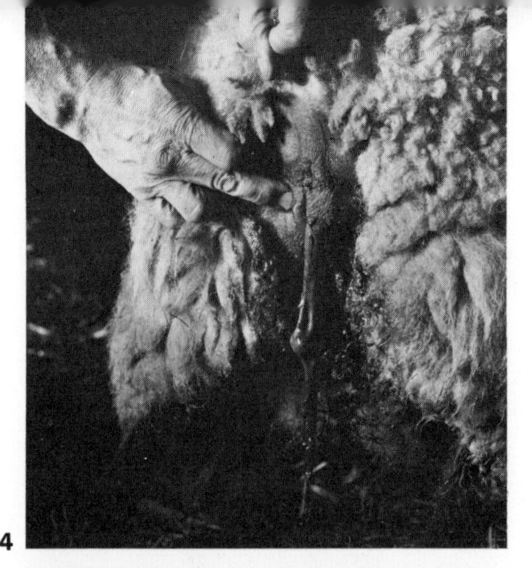

4

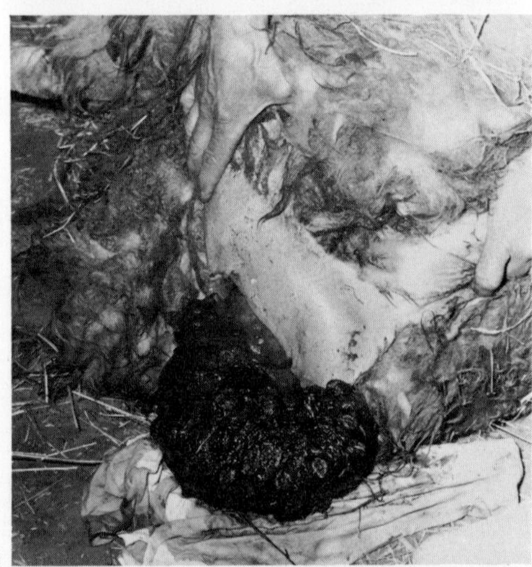

5

6

7

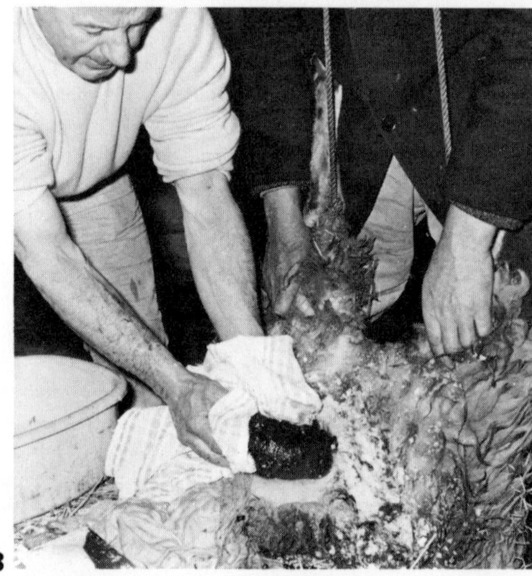

8

3. *Retained afterbirth (photo 4)*
This is the usual cause when the prolapse occurs several days after lambing.

Symptoms
These are unmistakable. The prolapsed uterus is studded with cotyledons or 'roses' with or without the afterbirth attached. As in the cow, it can be described as a 'bloody mess' *(photo 5)*.

Treatment
This is definitely a job for the veterinary surgeon.

The ewe is placed in the position for lambing and cervical prolapse return *(photo 6)*. The afterbirth is removed if necessary; the entire uterus is washed very carefully with hot water containing a mild concentration of a non-irritant antiseptic *(photo 7)*, and dried with a clean towel *(photo 8)*. This is a useful hint since the

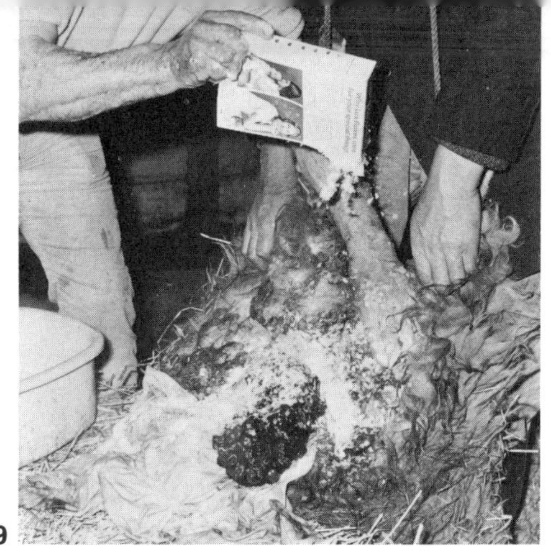

9

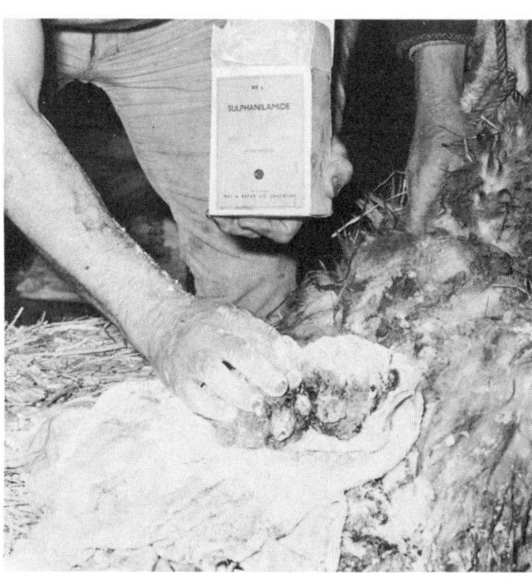

10

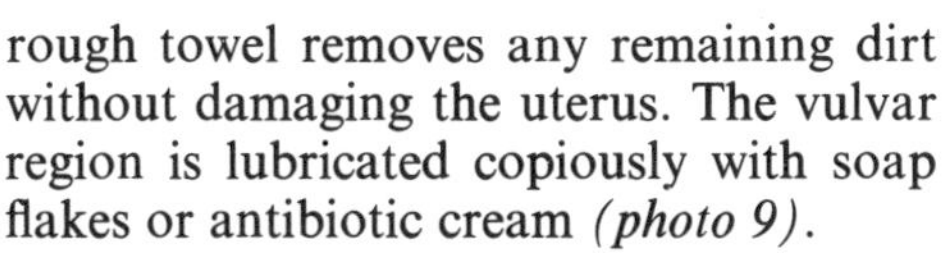

rough towel removes any remaining dirt without damaging the uterus. The vulvar region is lubricated copiously with soap flakes or antibiotic cream *(photo 9)*.

Dry sulphanilamide powder is massaged lightly over the entire uterine surface *(photo 10)*. The hands and arms are now thoroughly washed and copiously lubricated *(photo 11)*. The uterus is enclosed in the palms of the hands *(photo 12)* and, very gently, is replaced between the ewe's strains.

When back in position, it is essential to make sure that the tips of the uterine horns are fully turned backwards *(photo 13)*.

The vulva is then sutured in the same way as for cervical prolapse *(photo 14)*.

After-Treatment

If not given before the prolapse is returned, 100 cc of a 20 per cent solution of

11

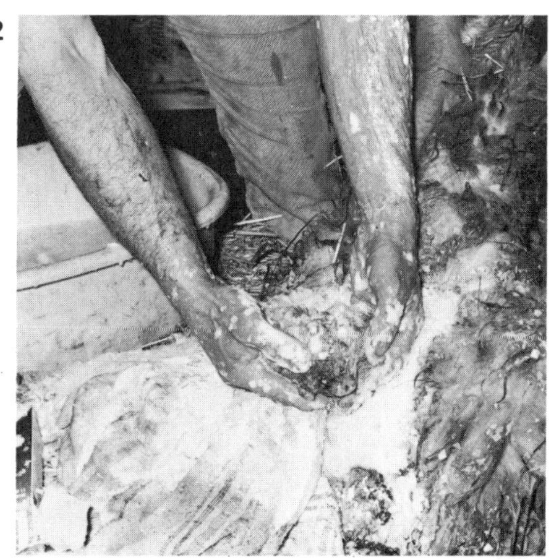

12

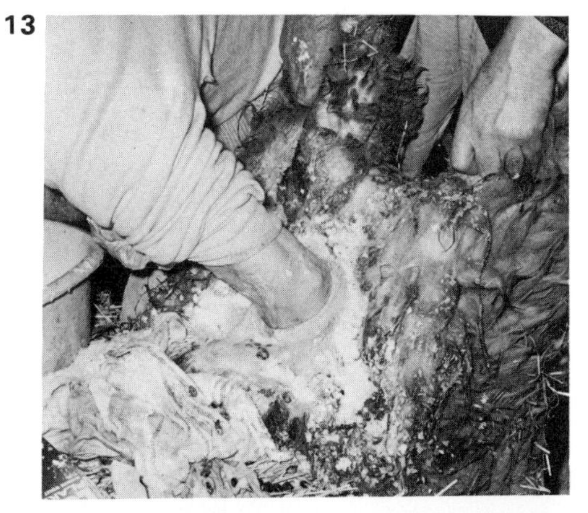

13

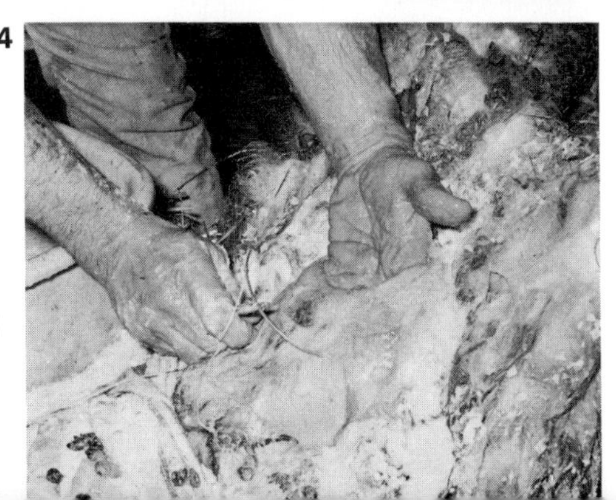

14

calcium borogluconate should be injected subcutaneously and 6 cc of penicillin or streptomycin given intramuscularly *(photo 15)*.

The antibiotic injections should be continued daily for at least 5 days.

Do The Ewes Recover?

With the modern antibiotic therapy the majority survive though, as with the cervical prolapses, it is wise to mark the ewes and cull them before the next breeding season.

The spontaneous type of prolapsed uteri and that associated with lambing sickness cannot be prevented, but that due to retained afterbirth can—by daily injections of penicillin or streptomycin until the afterbirth is voided.

In fact all cases of retained afterbirth in ewes should be treated thus as a routine. If not, septicaemia is an even greater hazard than prolapse.

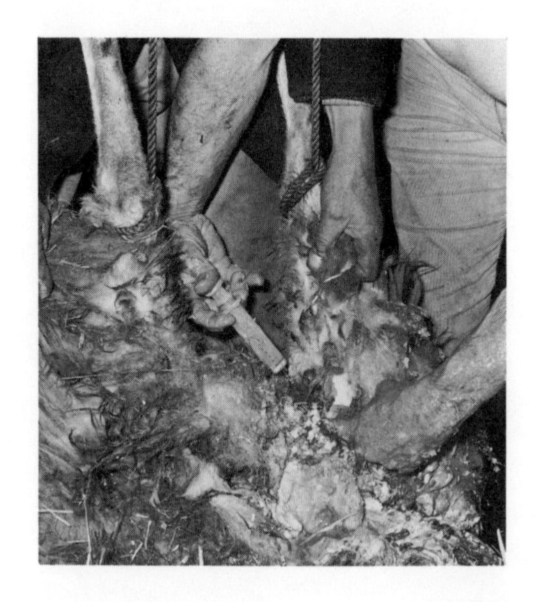

38
Malpresentations

WHEN THE two forelegs and head of the lamb are not coming first, then the shepherd is dealing with a malpresentation *(photo 1)*. If, as in most cases, the shepherd's hand is a large one, and provided professional help is readily available, it will no doubt pay handsomely to call in the veterinary surgeon.

For the remote sheep farmer, and for the shepherd with a small hand, a gentle touch and a great deal of experience, the following professional hints should be of considerable value.

Never attempt to correct a malpresentation with any part of the lamb in the passage. If possible—and this applies in the vast majority of cases—always, using copious lubrication, gently replace the impaction into the womb, where there is much more room to move about and where the danger of inflicting damage to the mother is correspondingly lessened *(photo 2)*. The ewe of course should be placed and kept in the upended position, until the presentation is corrected.

1

2

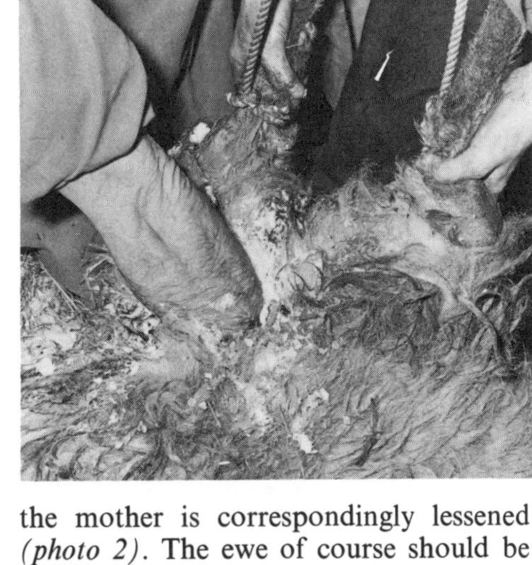

Common Malpresentations

HEAD BACK

Both forelegs are usually in the passage, but the head is turned right back into the uterus. The correct procedure is illustrated on the model *(photo 3)* and is as follows:

3

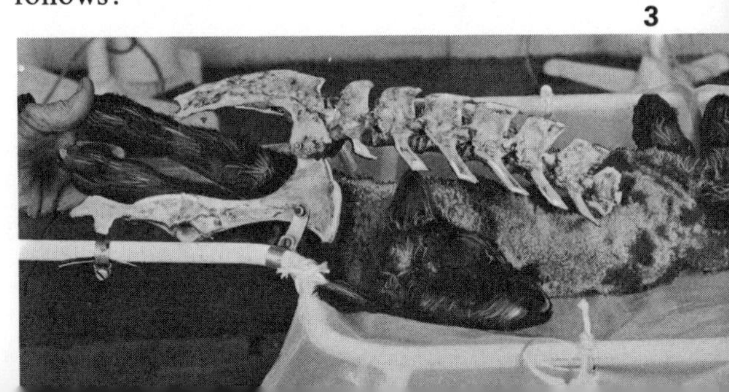

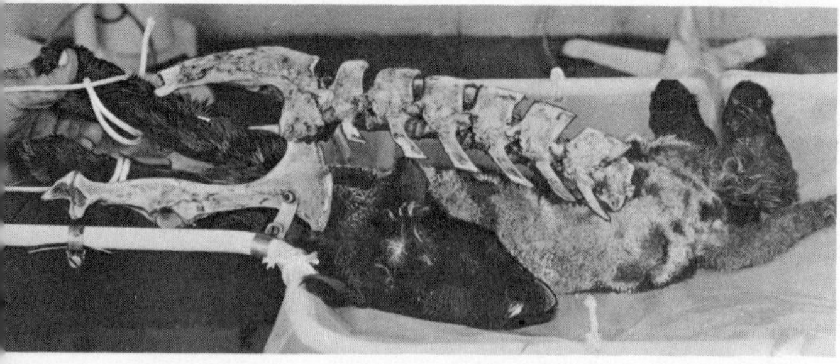

4

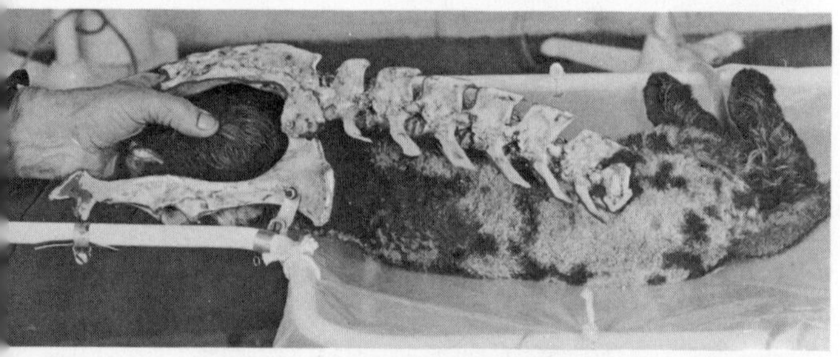

5

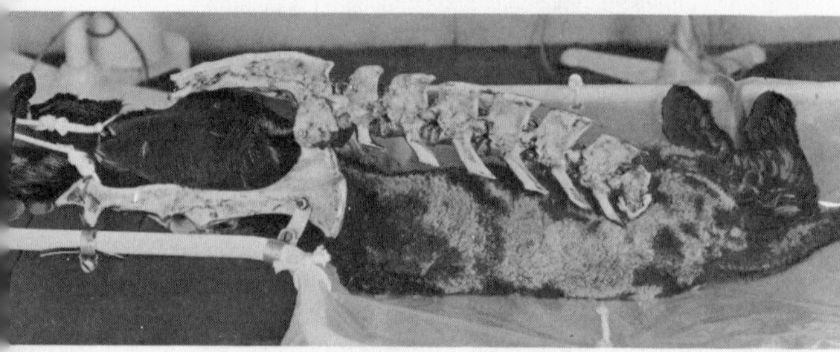

6

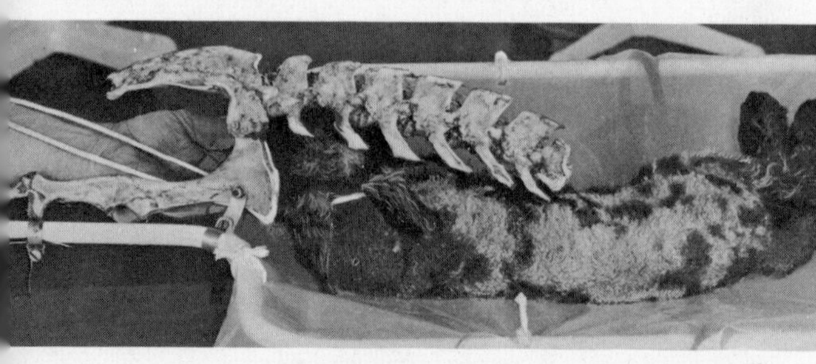

7

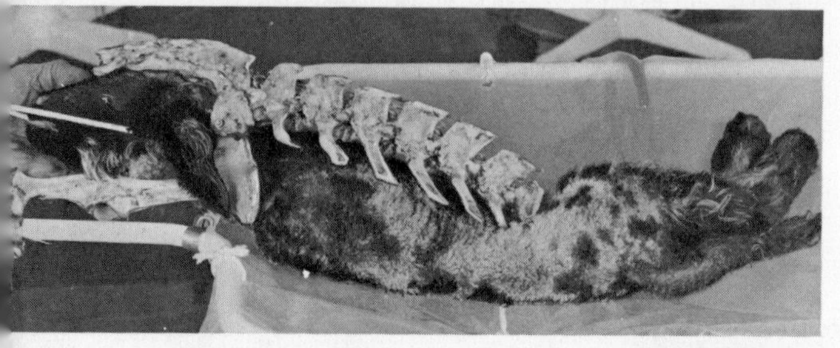

8

First of all, secure both forefeet with two lengths of strong white cord or sterile string *(photo 4)*. If the cord or string hasn't been boiled, then at least it should be coated with antibiotic cream. Then very gently, and using lots of soap flakes and warm water, push the lamb—between the ewe's strains—back into the womb. Grasp the head and turn it, and, with the finger and thumb in the eye sockets, ease it forward into the anterior vagina *(photo 5)*.

Now bring the legs forward one at a time *(photo 6)*, and as soon as the lamb is presented correctly, lower the ewe on to her side so that she can give the maximum of assistance with her straining. Proceed to help in the delivery, pulling as the ewe strains.

Whatever you do, don't attempt to pull the lamb's head forward in the palm of your hand; there won't be enough room for your hand and the head.

A good tip is to pass the centre part of another piece of cord over the top of the lamb's head to behind the ears *(photo 7)*, and use this to lever the head into the anterior vagina. Never cross the ends of the cord and never loop it round the neck. Exert the pressure only on the crest at the back of the head *(photo 8)*.

The basic principle for the correction of all malpresentations is the same as that just described and illustrated.

TWO LAMBS COMING TOGETHER

This can provide a puzzling tangle. There may be two heads, and one foot, of each lamb *(photo 9)*.

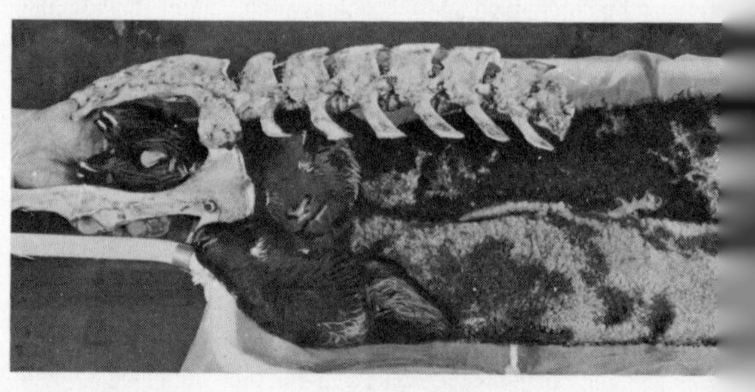

9

My advice for correction is this: with a length of sterile cord in the hand and using abundant lubrication, slowly replace the tangle in the uterus exerting pressure between the ewe's strains.

Select the nearer of the two heads. Pass the centre of the cord over the top of the head to just behind the ears, bringing the free ends of the cord outside the vulva *(photo 10)*. Maintain the cord in position by exerting moderate pressure. It is important to keep up this moderate pressure all the while with one hand, while working with the other, otherwise the string will slip.

Renewing the lubrication, pass the free hand on to the fixed head and from there run down the side of the neck to one shoulder, keeping the fingers hard against the lamb, then to the elbow, knee, and finally the foot. Bring the foot carefully upward and forward into the anterior vagina *(photo 11)*; this can be difficult and should certainly never be done at all hurriedly. *Use care, patience, lubrication and time.*

Now fix this foot above the fetlock with another piece of sterile cord *(photo 12)*. Manipulating the cord inside can be quite tricky, but with a little practice any shepherd with a reasonably small hand can soon become expert.

Once again renew the lubrication with lashings of soap flakes, and from the tied foot feel again for the knee, elbow and shoulder (keeping the fingers all the while hard against the lamb); up to the neck, to the head, and then down to the opposite shoulder, elbow, knee and foot *(photo 13)*. Bring the second foot up and forward, and secure this also with another sterile cord, again above the fetlock.

Next move the hand across to the other head, and push it gently back into the womb, at the same time keeping a steady pressure on all three pieces of cord *(photo 14)*.

Now bring one leg at a time forward, *and having placed the ewe on her side*, patiently and slowly assist in the delivery,

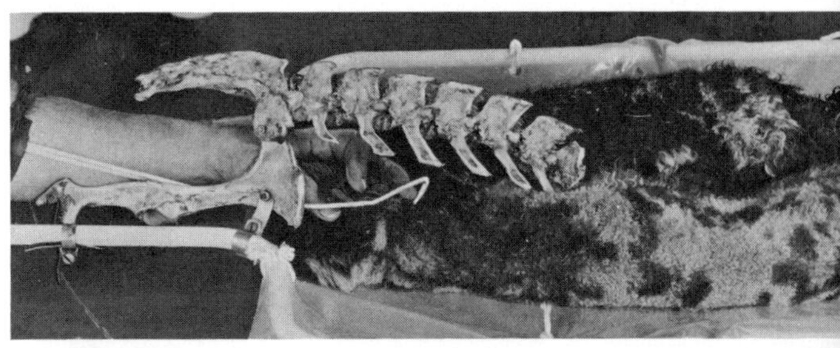

10

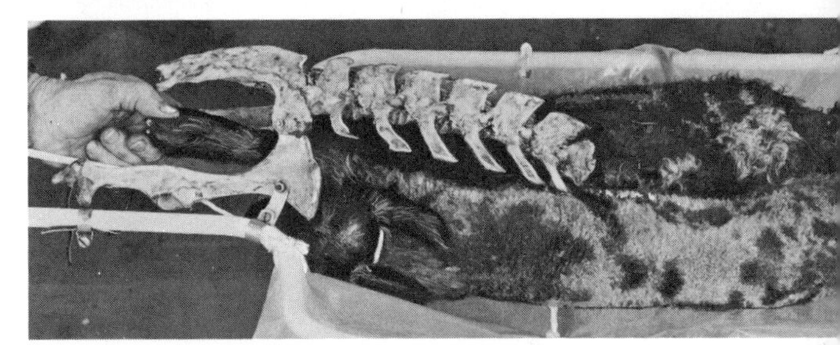

11

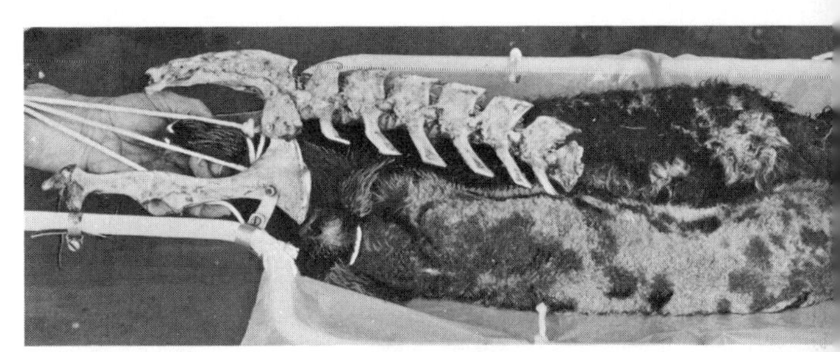

12

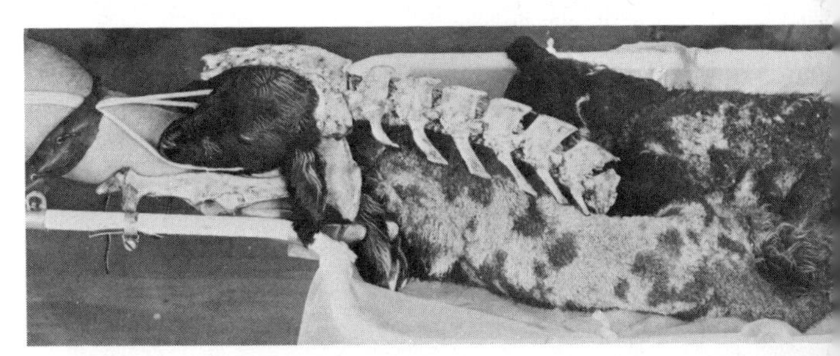

13

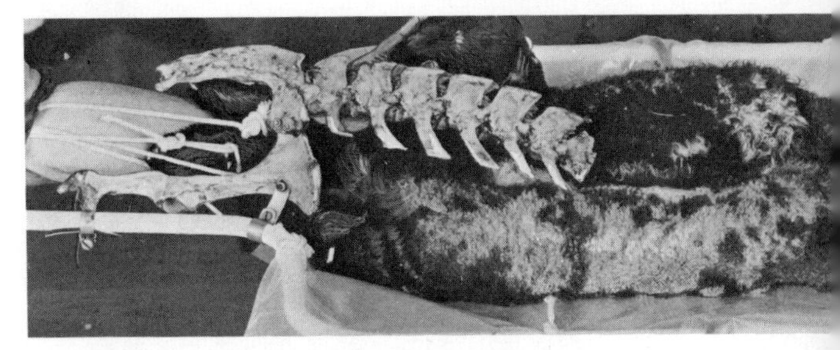

14

15

easing the head through the vulva and allowing the ewe to do most of the work *(photo 15)*. The second lamb, of course, will present no problems.

SINGLE HEAD PRESENTED WITH ONE OR BOTH LEGS BACK

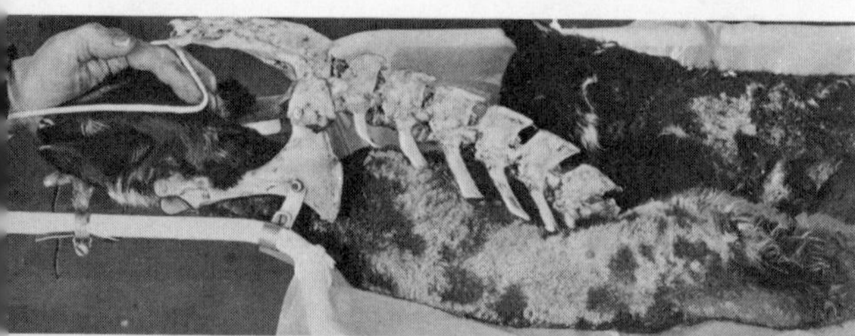

16

Follow the identical procedure just described as there is often another lamb behind *(photo 16)*, and you may well bring forward a wrong leg. As a matter of fact, this is the mistake that has most often been made when I'm called in.

HEAD AND ONE LEG OUTSIDE THE VULVA

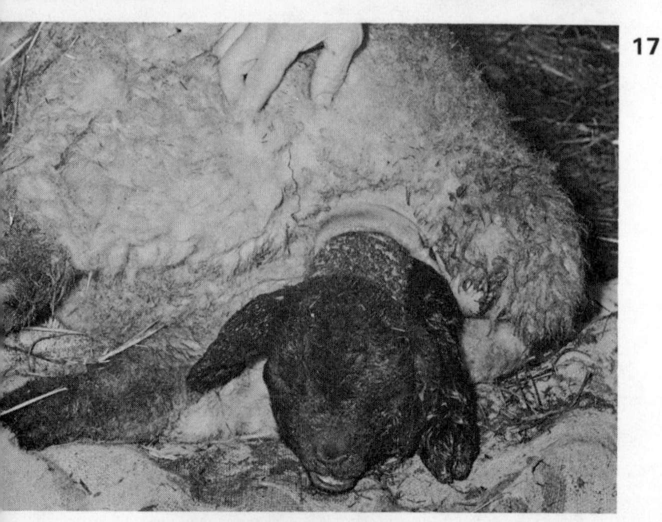

17

The correct procedure is, first of all, to place the ewe on her side and apply gentle pressure on the lamb, assisting the patient when she strains *(photo 17)*. I have found that approximately 80 per cent of lambs can be delivered with one shoulder back without damaging the ewe.

If the lamb won't move with gentle pressure, however, it is wrong to pull too hard. The ewe should then be held in the position for malpresentation correction, the vulva and protruding portion of the lamb lubricated thoroughly with warm water and soap flakes, and the lamb gently replaced into the uterus, exerting pressure, only between the ewe's strains *(photo 18)*. Such a replacement may take some time, but with a large single lamb it is well worth the time and patience. Once back in the uterus it is usually fairly easy to bring forward the missing leg *(photo 19)*.

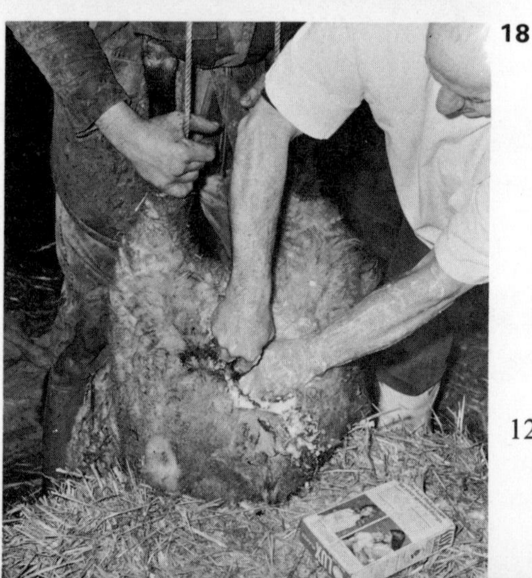

18

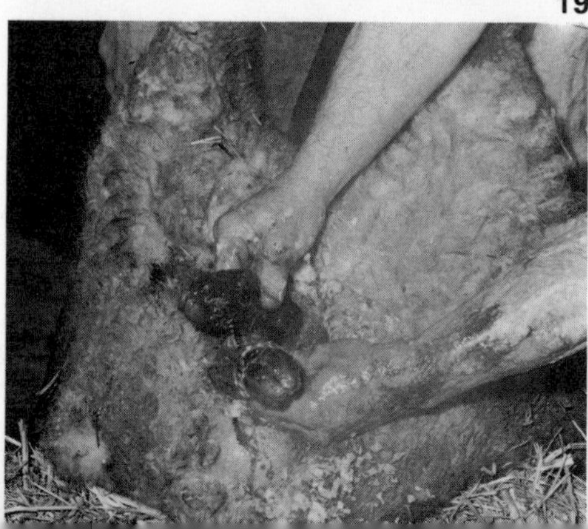

19

HEAD OUT AND SWOLLEN

When the head itself is out and swollen, and both legs are back, then delivery in one piece is difficult and should never be attempted unless the lamb is alive *(photo 20)*.

If alive, soap flakes provide probably the only and certainly the best key. The ewe should be placed in the suspended (or upended) position, and the lamb's head and inner ring of the surrounding vulva copiously lathered with warm water and soap flakes *(photo 21)*. Then, with the hand or hands cupped over the head, steady pressure should be exerted, increasing the pressure between each of the ewe's strains *(photo 22)*. Considerable patience is required here because the task may at first seem impossible. Gradually, however, vulva and vagina will relax and the lamb can be replaced at least sufficiently to bring one leg forward. Naturally, of course, it is always better to bring both forelegs forward, and I have found that this is nearly always possible in a ewe from the second crop onwards *(photo 23)*. Large first single lambs in this position present a special problem and these should be left to the veterinary surgeon.

If the lamb is dead and stuck with the head or head and one leg out, then the correct procedure is to cut the head off *(photos 24 and 25)*, then upend the ewe

20

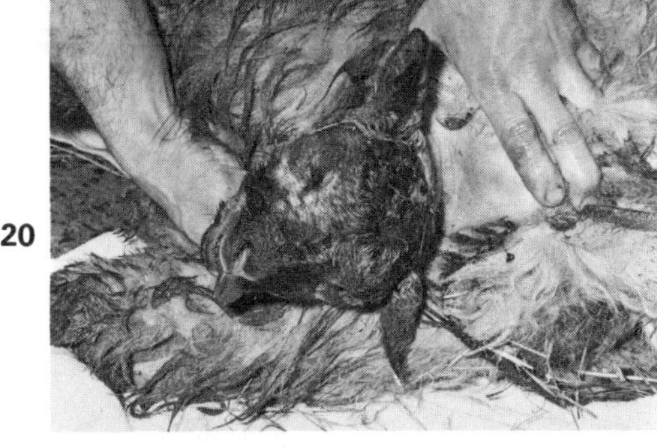

21

22

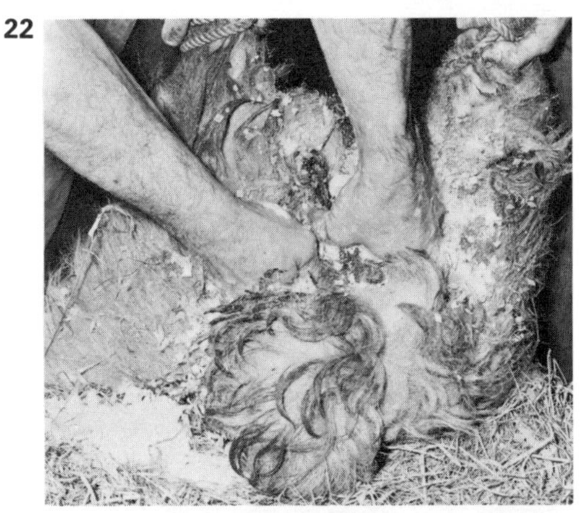

23

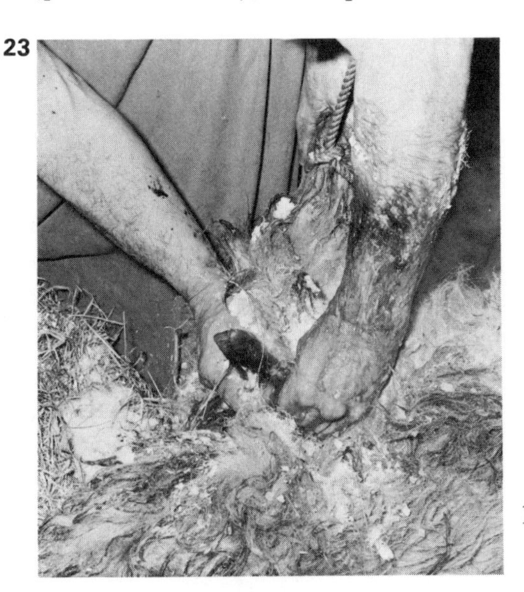

24

25

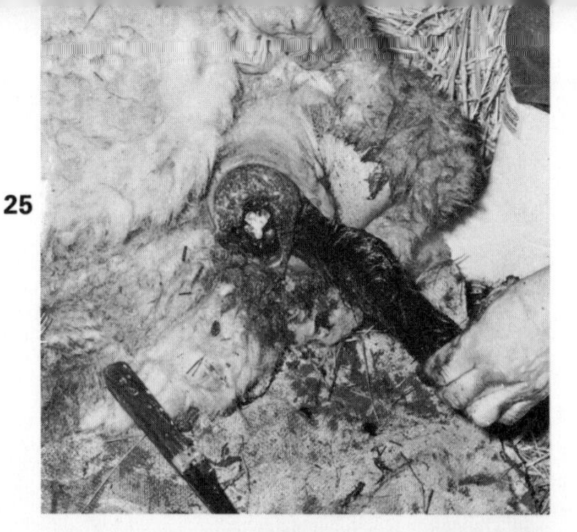

26

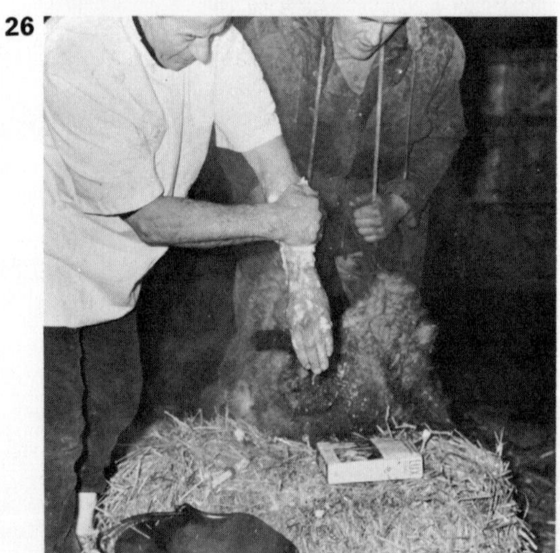

27

28

and replace the stump slowly, between the ewe's strains, using once again copious lubrication *(photo 26)*. When the foetus is back in the uterus, bring the forelegs forward one at a time; then replace the ewe on her side, and holding the fingers over the stump to protect the passage from laceration, deliver the lamb, with the ewe doing most of the work by herself *(photo 27)*. After the delivery, as with all bad lambing cases, inject the patient with an adequate dose of antibiotic *(photo 28)*.

POSTERIOR PRESENTATION

In this, the hind-feet of the lamb are coming first. They are identified by the feet being upside down *(photo 29)*, though it is always best to examine carefully to make sure the hocks, hind-quarters and tail are there *(photo 30)*, as occasionally the lamb is upside down.

29

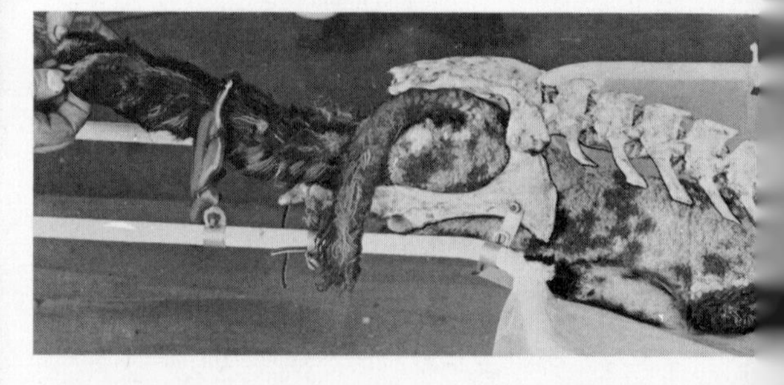

30

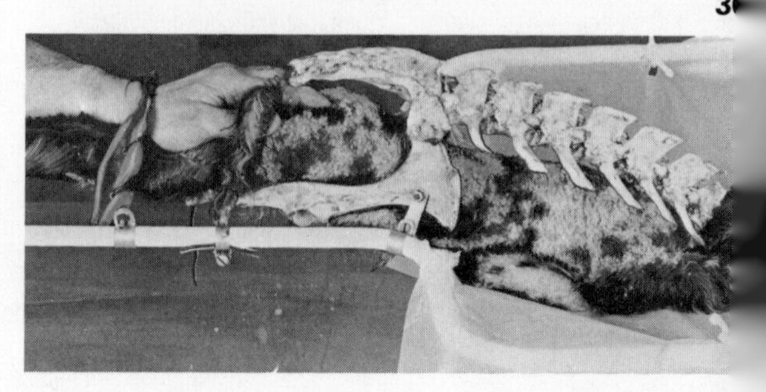

The correct technique is to lay the ewe on her side and assist by pulling gently on the hind-legs as the ewe strains *(photo 31)*, again allowing the ewe to do most of the work. However, with the posterior presentation, there is one very important point to remember: As soon as the entire tail is out of the vulva, the lamb's life is in danger simply because at that stage the navel cord, which provides the vital oxygen to keep the lamb alive, is stretched to the absolute limit and is compressed between the belly of the lamb and the ewe's pelvis *(photo 32)*.

The hints on how to get a live posterior lamb, therefore, are: first of all, to rotate the lamb through about one-quarter of a circle during the final withdrawal, and secondly, to make the final pull a rapid one *(photo 33)*. Any delay with the navel blood supply cut off will produce a dead lamb. The reason for this is that, once the navel cord breaks, the lamb has to use its own lungs to provide the oxygen, and if the head is still in the ewe when the first deep breath occurs, then the lungs are filled with mucus which suffocates the lamb. A rapid rotating delivery results in a live lamb *(photo 34)*.

THE BREECH PRESENTATION

When the hind-quarters are coming first with the tail often showing outside the vulva, the presentation is described as a 'breech' *(photo 35)*.

31

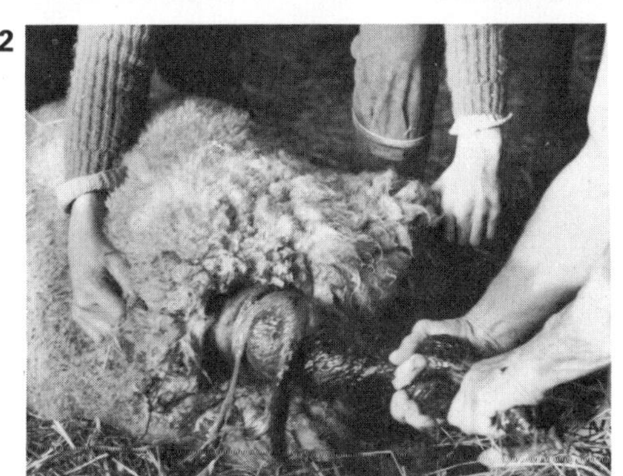

32

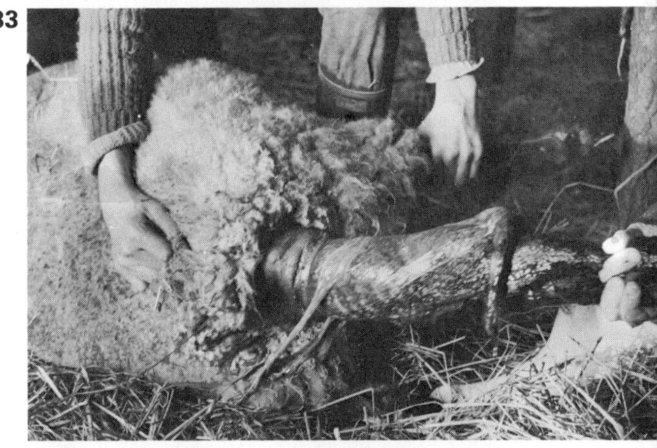

33

34

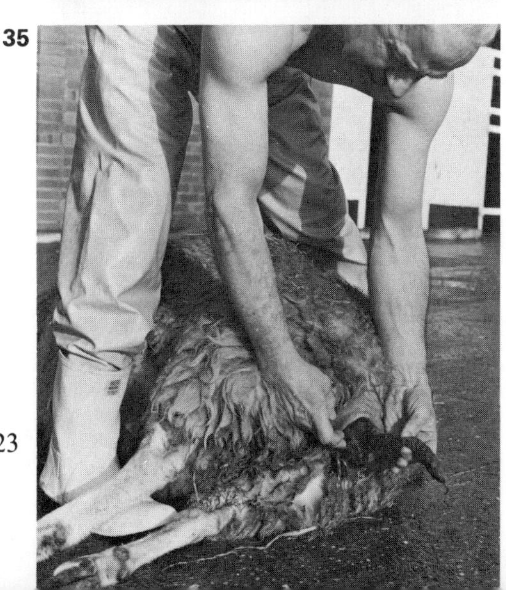

35

36

Upend the ewe, fill the vagina with soap flakes *(photo 36)* and replace the lamb slowly between the ewe's strains *(photo 37)*. Never try to force it back or the feet will rupture the womb. Now run the hand from the tail down to one hock; push the hock upwards and backwards, and the foot will come towards the vagina. Grasp the foot and bring it forward *(photo 38)*.

Repeat on the other side (from tail to hock), hock upwards and backwards, and bring the second foot forward *(photo 39)*. Replace the ewe on her side and assist delivery exactly as for the posterior presentation (which, of course, it now is), remembering to turn the lamb partially to ease the pressure on the navel cord and to make the final withdrawal as speedy as possible *(photo 40)*.

37

39

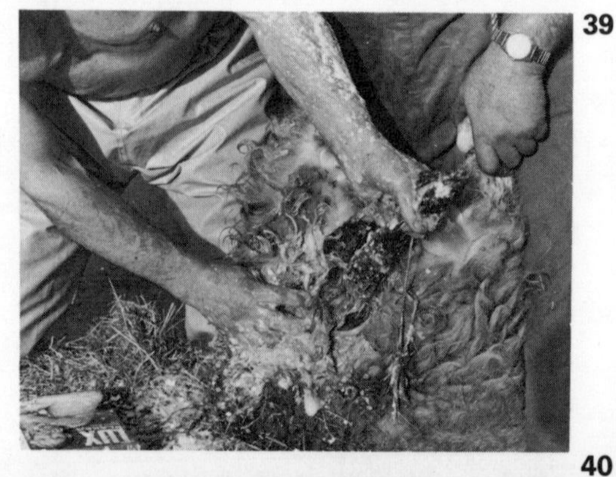

38

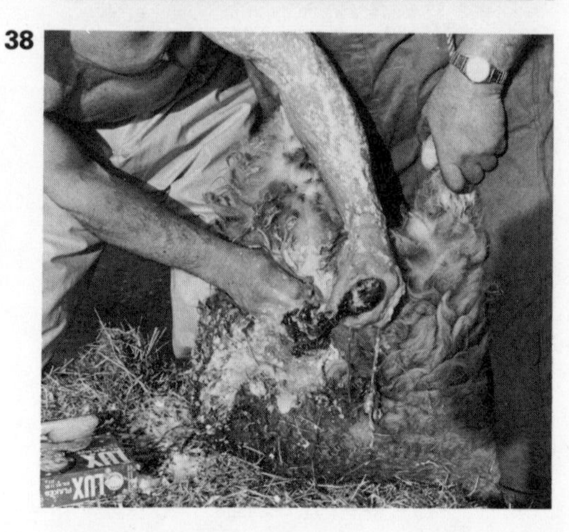

40

TORSION OF THE UTERUS

Just occasionally in the pregnant ewe the body of the uterus becomes twisted on its own axis *(photo 41)*. The twist or torsion may be partial or complete.

When the ewe comes to full term, she will give every sign of first-stage labour but will not get on with the job.

Examination per vagina will reveal either a complete blockage of the passage or a corkscrew entry to the cervix or neck of the uterus.

Torsion can be easily dealt with provided it can be spotted in reasonable time. If the shepherd has missed the early labour signs, the lambs will die and putrefaction will set in.

The correct procedure is to insert your hand into the vagina as far as the twist. Then get an assistant or assistants to roll the ewe first one way and then the other *(photo 42)*. When the rolling is in the correct position, the twist will be felt disentangling itself and the way to the cervix will become clear and normal.

Usually the cervix or entrance to the womb is not fully opened after the torsion correction. Muscle relaxants and antibiotics then have to be injected, and the ewe should be left for at least 12 hours or until the cervix is fully dilated.

If the lambs are alive and healthy, normal birth usually follows the torsion correction.

THE DOG-SITTING POSITION

A common presentation, and an extremely difficult one to cope with, is when the centre part of the lamb's spine is coming first, with the head and all forelegs pointing into the uterus *(photo 43)*.

This is one for the veterinary surgeon unless the shepherd is very experienced and has a comparatively small hand.

The correct procedure is, first of all, to get the ewe into the suspended position. Fill the vagina with soap flakes, thoroughly lubricate the hand and arm then introduce it very carefully. Keeping the

41

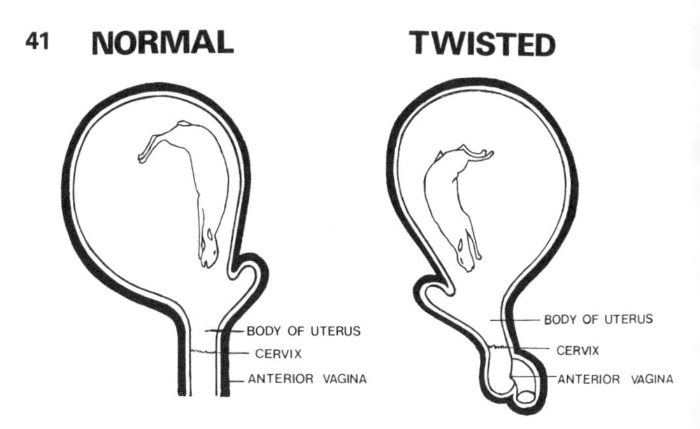

42

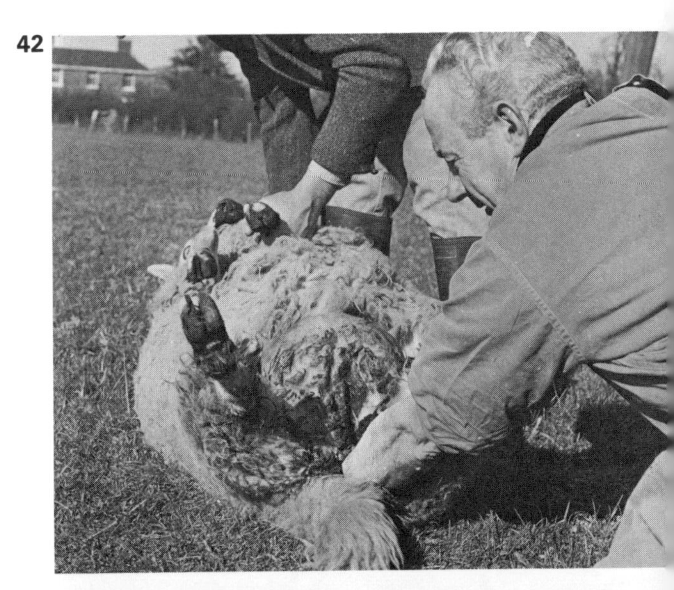

43

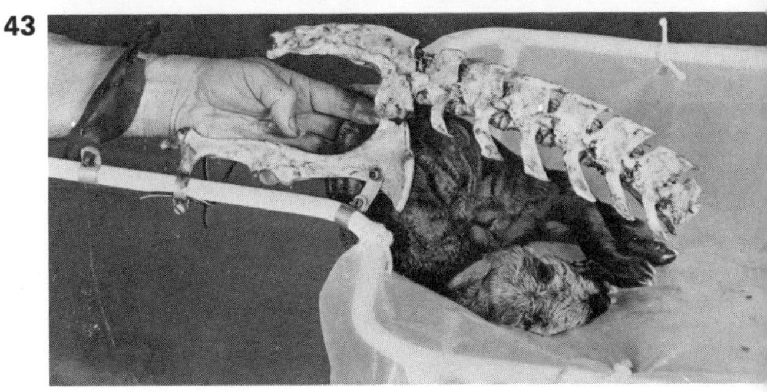

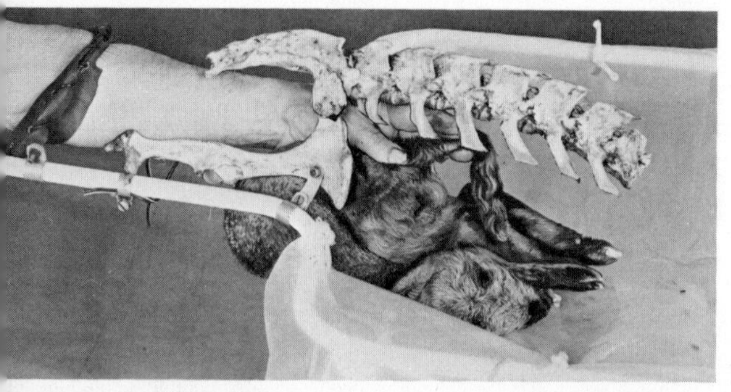

44

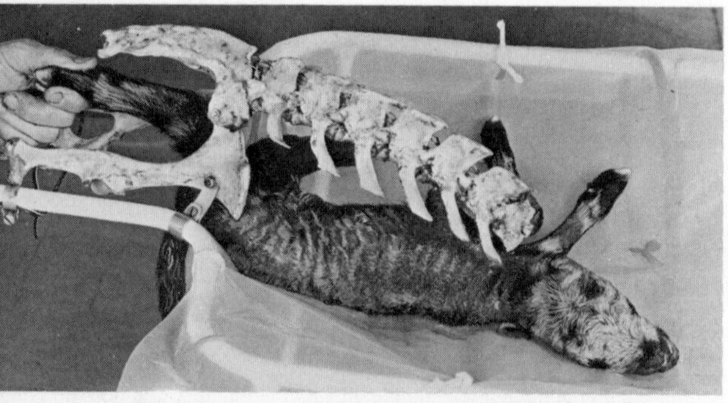

45

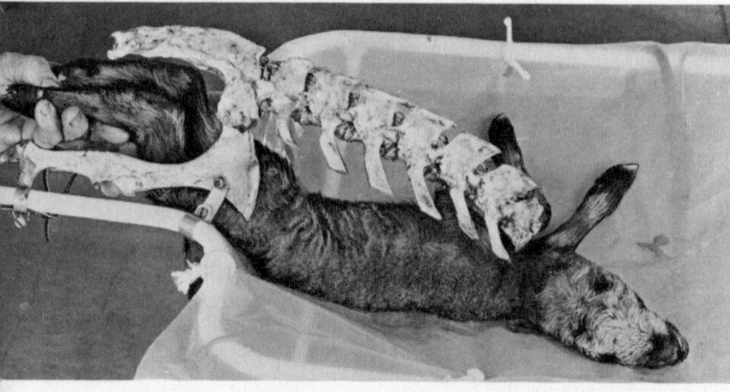

46

fingers pressed hard against the lamb, search patiently for the lamb's tail *(photo 44)*. I have found, over the years, that a dog-sitter is always easier sorted out from the hind end.

From the tail run the hand down one hind-leg to the hock; then push the hock gently upwards and backwards and bring forward the first hind-leg *(photo 45)*.

47

Follow in again along the leg to the tail, then over to the other side to search for the other hock, always keeping the hand hard against the lamb. Bring forward the other hind-leg *(photo 46)*. Lay the ewe on her side and deliver the lamb as in a straightforward posterior presentation.

To sum up obstetrics in the ewe: always give Nature a fair chance and leave the ewe for a minimum of 4 hours. When examining, get the ewe into the correct upended position and use cleanliness, lubrication and careful manipulation.

In malpresentation, secure identifiable parts and, using ample lubrication, gently replace the lamb in the womb before correcting, always moving gently and carefully and keeping the fingers hard against the selected lamb while looking for a leg or a head.

Never panic into rough groping around: patience, combined with the three golden rules of cleanliness, lubrication and careful manipulation, will resolve the vast majority of ovine obstetrical problems *(photo 47)*.

39
Foetal Abnormalities

FOETAL ABNORMALITIES (i.e. malformed lambs at birth) are fortunately comparatively uncommon in sheep. Perhaps the commonest is the dropsical lamb—illustrated here. This condition, known in some areas as 'water-belly', has been attributed to the feeding of pregnant ewes on roots. This is not the case, though obviously it is unwise to feed excess turnips, mangolds or potatoes to pregnant, or for that matter to any sheep, because of the dangers of poisoning. Mangold and potato poisoning can be just as serious as kale poisoning.

40

Retained Afterbirth

THE PASSING of the afterbirth in the ewe is the third stage in the natural normal birth. Fortunately the ewe's afterbirth is not often retained, but when it is there is always a reason.

Cause

1. Premature birth *(photo 1)*. The lambs coming perhaps only a few days before their time and appearing normal in every way.
2. Uterine fatigue after twins or triplets.
3. Calcium or magnesium deficiency. Both these minerals are concerned with the contraction of muscles, and a slight deficiency of either can prevent the uterine muscle contracting sufficiently to detach the afterbirth.
4. Uterine infection, for example, vibrosis or salmonella infection.

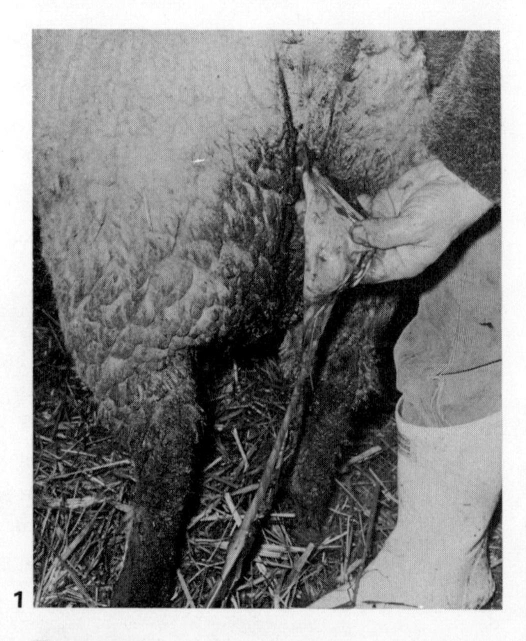

1

Treatment

Never attempt to remove a ewe's afterbirth manually. Manipulation of the uterus will lead to incessant straining and possibly prolapse.

The correct treatment is to maintain a systemic cover of antibiotic until the afterbirth drops. This can be done by daily injections of streptomycin or penicillin or by using a long-acting antibiotic *(photo 2)*. But your veterinary surgeon will advise you, and where necessary will no doubt supply you with the syringe and necessary drugs.

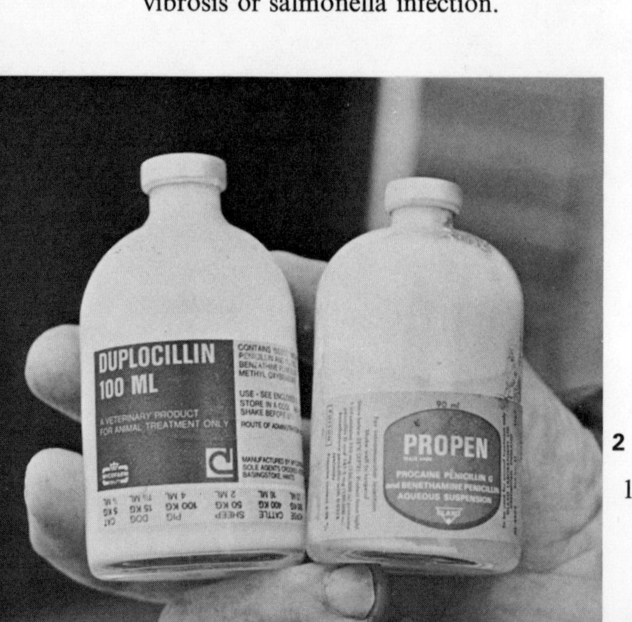

2

41 Mastitis

MASTITIS SIMPLY means inflammation of the udder *(photo 1)*. Obviously, therefore, it occurs in ewes and usually after lambing.

1

Cause

Nearly always mastitis in the ewe is caused by a very powerful germ called the *Staphylococcus aureus*.

Predisposing Causes

Scratches, wounds or infection on the skin of the udder and teats. The ulcers secondary to orf (see Orf, p. 155) provide an ideal field for the growth of the staphylococcus.

The germ gets into the udder either via the damaged skin surface of the udder or up the teat canal.

Symptoms

The hyperacute type produces the gangrenous mastitis or 'black garget'. The ewe runs a high temperature, goes off her food completely and stands with hind-legs wide apart and backwards *(photo 2)*. The affected quarter is hot, swollen and painful, but towards the teat it is ice-cold, black or blue and gangrenous *(photo 3)*. If you draw the affected teat, a watery blood-stained fluid with a peculiar sweet

2

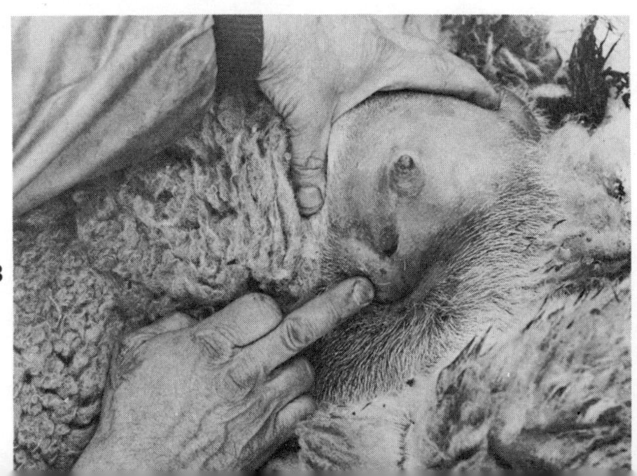

3

bread smell is emitted *(photo 4)*.

I have found that such cases may die in the first 24 hours. If they live after that, the body temperature returns to normal and the ewe starts to eat again. When this happens, the gangrenous quarter sloughs or drops off during the following month or 6 weeks.

In less acute cases gangrene does not develop, the affected quarter being merely hard, hot and swollen.

In chronic cases *(photo 5)* the quarter becomes 'indurated', *i.e.* hard. This is due to the milk secreting tissues being destroyed and replaced by hard fibrous tissues.

Treatment

The acute gangrenous case should be isolated immediately, to avoid spread of the infection, and should be injected with maximum doses of antibiotic *(photo 6)* in an attempt to save the ewe's life. Even though the affected quarter subsequently drops off, the ewe should still be able to suckle a single lamb on the remaining teat.

The less acute cases may respond completely to a course of broad-spectrum antibiotic. I have found the most effective antibiotic to be chloramphenicol containing cortisone and antihistamine *(photo 7)*.

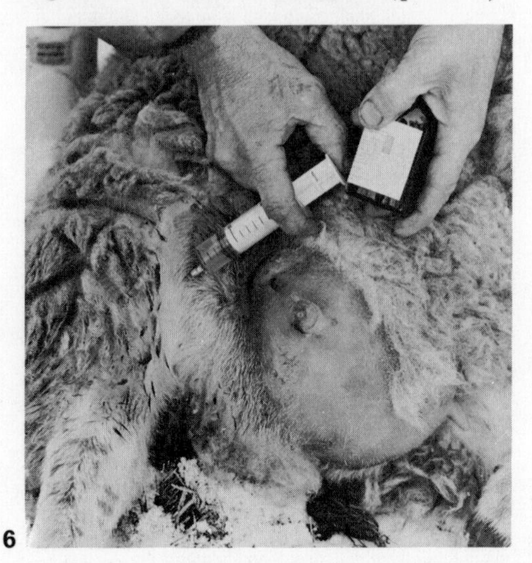

6

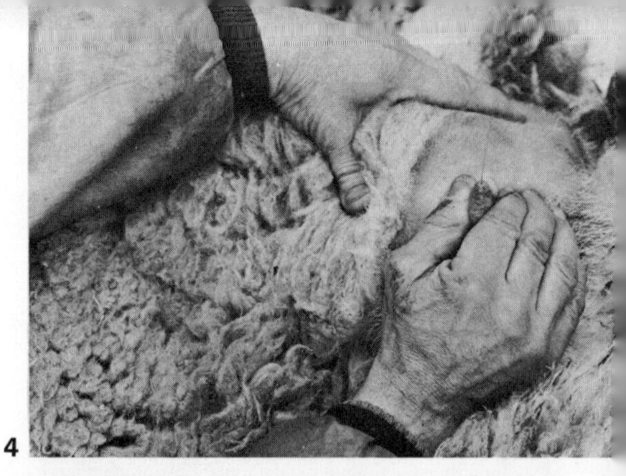

4

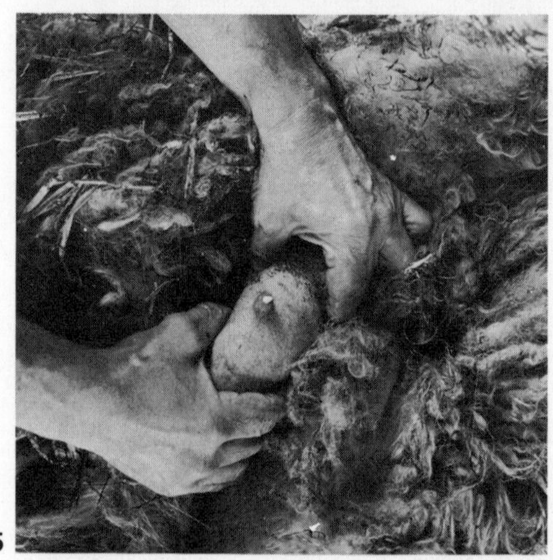

5

The chronic cases are not worth treating and should be marked for culling.

Control Or Prevention

Examine the udders of all ewes before buying them. A lumpy quarter denotes a potential carrier.

All clinical cases should be marked and culled.

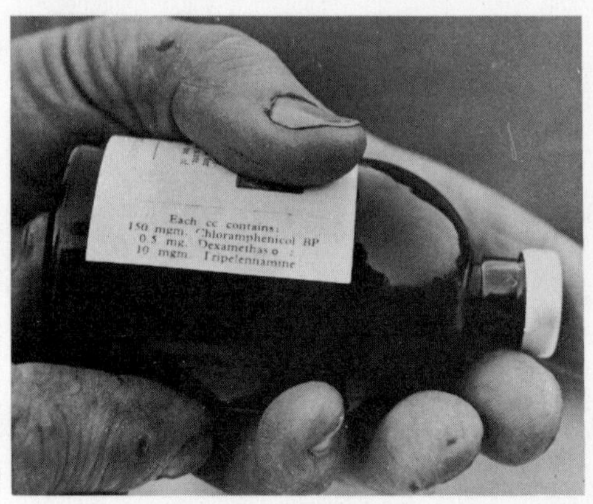

7

42

Care of the New-Born Lamb

OFTEN AN apparently healthy new-born lamb may appear to be dead *(photo 1)*. In such a case the heart is nearly always beating, and all that is required is the stimulation or induction of the first breath.

Respiratory stimulants are now available from your veterinary surgeon *(photo 2)* ; a few drops on the back of the tongue and the lamb will snatch its first breath.

Blowing down the throat and tickling the nostril with a piece of straw, plus artificial respiration, are all valuable aids to getting the lamb going. However, I have found the most useful hint of all is to literally swing the lamb round several times holding it by the hind-legs—above the hocks, *i.e.* like swinging the proverbial cat *(photo 3)*. Over the years I must have saved the lives of thousands of lambs by this simple method. The greatest advantage is that it can be done immediately,

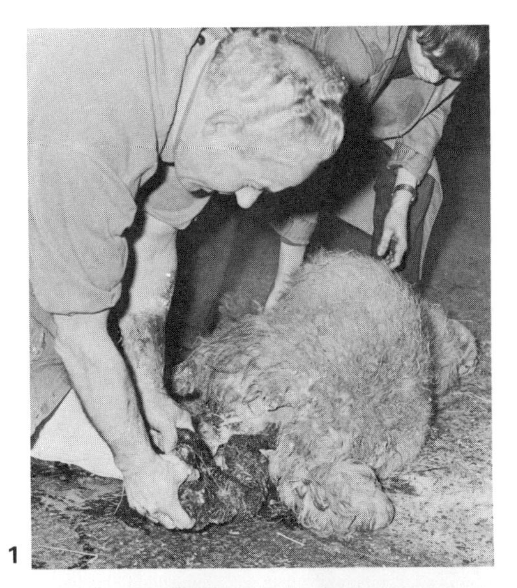

1

2

3

4

5

thus avoiding loss of valuable time.

Reasonable warmth is an important factor in the survival of many lambs. In lowland flocks, therefore, and wherever possible in the hills, lambing should be planned in sheltered fields or in covered lambing pens.

If the lamb is very feeble, a teaspoonful of gin or brandy *(photo 4)* and a night in a warm box by the kitchen fire will work wonders.

Once the lamb has filled his belly with his mother's milk, he rarely looks back. For this reason it is vital to check the udder of every ewe immediately after lambing, to make sure that the quarters are functioning correctly *(photo 5)*. Many lambs die from starvation simply because the teats of the mother are blocked or damaged.

If lamb dysentery is a special problem on the farm, then the new-born lambs should be injected with a booster dose of concentrated lamb dysentery serum. If despite this, diarrhoea should appear, then veterinary advice should be sought at once (see *E. coli* Infection, page 54).

43

Persuading a Ewe to Adopt an Orphan Lamb

ONE OF the most difficult tasks in any flock is to get a ewe to take to a lamb that is not her own.

Several traditional methods are repeatedly tried, with only moderate success. During the course of my practice, I have discovered a simple and apparently infallible method, which I would like to pass on to all shepherds.

Put the orphan lamb with the ewe in a pen or loose box, and bring the dog in. The ewe's in-born maternal instinct will more or less immediately stimulate her to offer protection and refreshment to the lamb. Keep the dog in the pen for 3 or 4 minutes.

Repeat this twice daily until the ewe has settled down with the lamb. Seldom does this take longer than 4 or 5 days at the most.

44
Docking and Castration

I AM convinced that both these practices will largely be discontinued when the ever-improving nutritional standards obviate the traditional necessity.

1

2

However, since both castration and docking are still widely practised, I must admit that the best and most humane method appears to be the application of the rubber rings as soon after birth as possible *(photo 1)*. Ideally, the lambs should be at least 1 day old and should be vigorous and active.

Two important points to remember: where rubber rings are used, it is more than ever vital that the ewe flock should be fully vaccinated against tetanus with a booster dose immediately prior to lambing. The skin wounds produced by rubber rings can provide ideal growing grounds for the tetanus bacillus *(photo 2)*.

Secondly, always make sure that both testicles are in the scrotum after the ring has been fitted *(photo 3)*.

3

GENERAL SHEEP DISEASES

45
Actinobacillosis

1

2

ACTINOBACILLOSIS AFFECTS the head region and shoulders of certain sheep in south Scotland. It is also seen in Sweden, Denmark and the USA.

Cause

The same germ that causes wooden tongue in cattle—the *Actinobacillus lignieresi*.

Predisposing Cause

Any wound that will allow the germ to enter and grow; for example, the wounds produced by rams fighting each other.

Symptoms

Small abscesses, one or several and about the size of a walnut, form on the lips, or neck or sometimes the shoulder *(photo 1)*.

If untreated, secondary abscesses can form, via the bloodstream, in the kidneys, lungs and liver. When this happens, the sheep rapidly loses condition.

Treatment

Using strict aseptic precautions, lance the superficial abscesses *(photo 2)* and inject the patient with antibiotic—streptomycin or penicillin—daily for 3 days.

Prevention

Since the source of infection is still not fully understood, there are at the moment no logical preventive methods.

Fortunately, only a few of the flock are likely to be affected.

46 Anthrax

AN ACUTELY fatal disease of sheep occurring throughout the entire world.

Cause

It is caused by a germ—the *Bacillus anthracis*.

In this country the germ is found in imported feeding stuffs and fertilisers such as bone meal.

Occasionally, the disease will flare up on a pasture in the vicinity of a previous case, or on land on which undiagnosed anthrax cases have been buried, perhaps 20 or even 30 years previously. The anthrax organism surrounds itself with a very strong capsule or spore and can survive in the soil for many years, rising to the surface on numerous occasions.

Symptoms

Death of one or several apparently healthy sheep. In suspect cases, a blood sample should be taken from an ear vein *(photo 1)* and examined microscopically, especially when a full preventive medicine plan is in operation.

1

Anthrax is a hyperacute disease causing a sudden high fever with the temperature rising to 108° F. The affected animal stops eating and is obviously desperately ill. It usually dies within a few hours.

Just occasionally, in the less acute form, the sheep may live for a day. If it does, it shows clear signs of acute abdominal pain and develops a bloody diarrhoea with the watery dung almost pure blood.

Treatment

If spotted early, the sheep may respond spectacularly to a massive dose of penicillin (three million units) given intramuscularly *(photo 2)*. Several times I have successfully treated affected sheep in flocks where deaths had occurred and where we were taking great care to check the tem-

2

peratures of any of the flock that were even slightly off colour.

Post-mortem Findings

It is illegal to open an anthrax carcase, but occasionally one may do so in error when suspecting hypomagnaesaemia, enterotoxaemia or even lightning stroke.

The typical signs of anthrax are multiple haemorrhages throughout the entire body, especially in the intestines. The spleen may be grossly enlarged. If such a picture presents itself, then the case should be reported to the police or to your veterinary surgeon immediately.

Prevention

Further cases can be prevented by changing the food completely or by cutting the concentrate ration by half.

For anthrax-infected farms, that is, where the disease is known to be endemic in the pastures, a vaccine can be obtained from the Animal Health Department of the Ministry of Agriculture, Fisheries and Food through your veterinary surgeon.

47
Contagious Ophthalmia

THIS DISEASE, known also as Pink Eye and Heather Blindness can affect the eyes of sheep of all ages *(photo 1)*. It is similar to New Forest Disease in cattle.

Cause
It is caused by a rickettsial organism called the *Rickettsia conjunctivae*.

Where The Rickettsia Comes From
The bug lurks in the eyes of many apparently normal 'carrier' sheep and becomes active only when there is some damage to the eye surface—damage caused by foreign matter such as dust particles, long grass, chaff, irritation by flies, etc. This fact probably explains why the disease is most common in the summer and why it often flares up during dry, windy weather.

What Happens
The damage allows the resident organism to get going. The Rickettsia pierces the surface of the centre part of the eye (called the cornea) and starts to multiply.

If untreated, it forms first of all a white pin-head which rapidly increases in size. If still untreated, a yellow pointing abscess may form which eventually ruptures, leaving a filthy raw ulcer which can take up to 3 or 4 months to heal.

During this time the Rickettsia in its most powerful form is present in all the eye discharge. The wind may blow such discharge over considerable distances, contaminating the surrounding pasture. This probably explains why the disease spreads rapidly and sometimes alarmingly through the flock. Of course, the strong Rickettsia is also carried from one to the other by flies.

1

A certain number of the recovered animals remain carriers and they re-infect the flock in subsequent years.

Symptoms
The first sign is a running eye. Both eyes may be affected and one or other may be closed.

If neglected, the eye will quickly film over and the sheep will be in considerable pain with consequent loss in condition *(photo 2)*.

If untreated, the eyes fill up with a yellowish pus and may ulcerate. The sheep become blind and tend to stay in one place, thus reducing their grazing area and contributing further to their loss in condition.

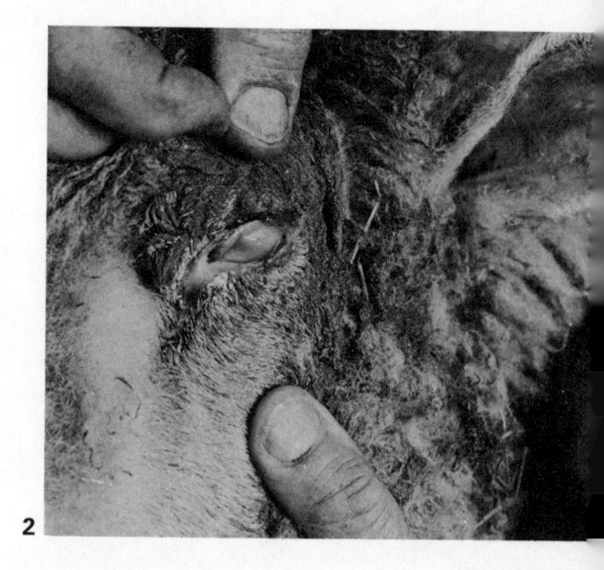
2

Treatment

Obviously, because of the rapid development and spread, it is very important to treat the disease at the earliest possible stage. There are a number of first-class preparations in the form of ointment, drops or emulsions. I have found that the most successful and easiest to use application is an emulsion containing chloramphenicol and dapsone *(photo 3)*. A single application inserted early on will often effect a cure within a few hours.

Even after the eye is 'marked' and clouded over, results can be quite spectacular, but the eye may have to be dressed daily for a considerable time.

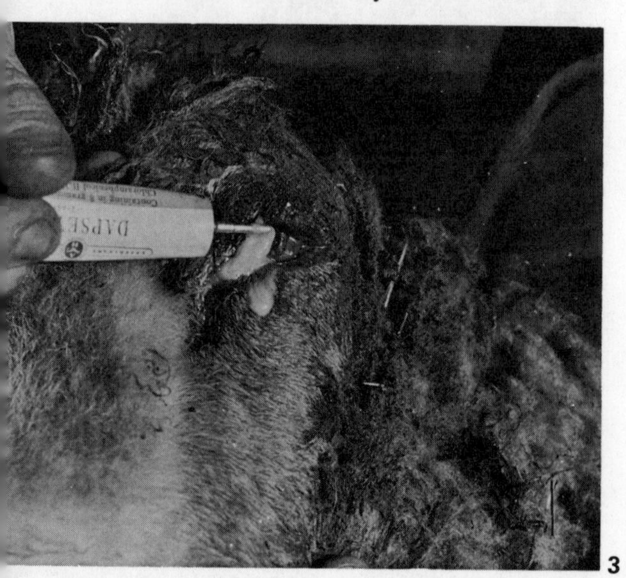
3

Can It Be Prevented?

Unfortunately no, but it can be controlled to some extent by a constant vigil and prompt treatment. I am sure that every sensible shepherd knows this and takes considerable care not to allow the spread. One obvious precaution is to isolate the affected sheep.

The reason why complete control is impossible is simply because of the persistence of apparently normal carrier animals after an outbreak.

48
External Parasites

MODERN DIPS have greatly reduced the economic importance of external parasites.

(a) LICE—PEDICULOSIS

Lice are common in sheep but are not considered of any great significance. They are kept in check by routine dipping against other external parasites.

The names of the common lice are of little interest to the sheep farmer but for the veterinary student there are three common species—two are illustrated in diagram A:

The biting louse—*Damalinia ovis*, which affects the woolled parts.

The sucking lice—1. *Linognathus ovillus*; 2. *Linognathus pedalis*. These are blue-black in colour and affect mainly the lower parts of the body and legs.

(b) PSOROPTIC AND SARCOPTIC MANGE *(SHEEP SCAB)*

Fortunately sheep scab has been eradicated from Britain by the regular routine use of the highly efficient modern dips. The disease is therefore of no economic interest to farmers.

Consequently, the following information is mainly for the benefit of veterinary and agricultural students.

Cause

It is caused by two mites called the

Diagram A

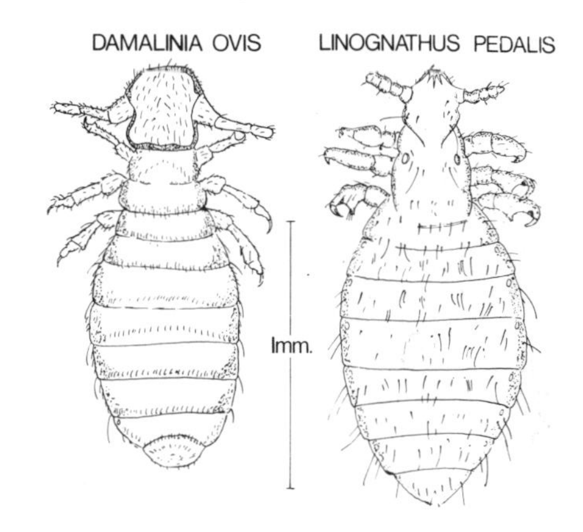

Psoroptes communis var ovis and the *Sarcoptes scabiei var ovis (see diagram B)*. When present in Britain, the sheep scab was worst in hill sheep because of the difficulties in inspecting them.

Diagram B

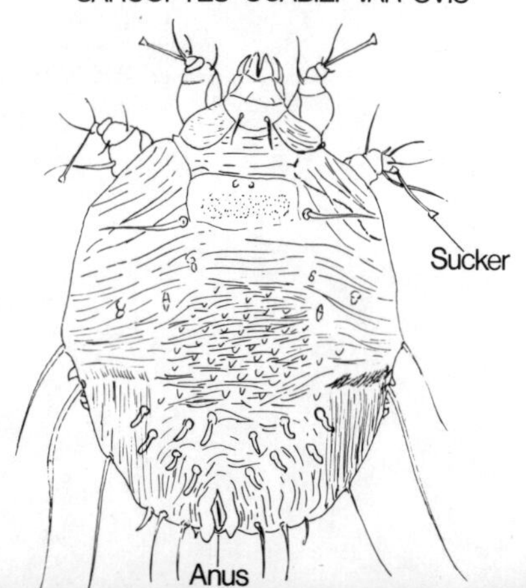

1

Symptoms
Sarcoptic mange affected those parts least covered by wool, forming a thick scab.

Psoroptic mange *(photo 1)* affected the woolled areas, especially along the back and sides towards the tail, causing loss of wool and scabs or crusts surrounded by moist or wet rings.

Needless to say, affected sheep were itchy and constantly rubbing themselves against posts, etc.

Treatment And Prevention
Sheep scab is a notifiable disease in Britain. Any suspect cases should therefore be reported to the police or to your veterinary surgeon.

2

Dips containing benzene hexachloride have been mainly responsible for the eradication of the sheep scab mites.

(c) CHORIOPTIC MANGE

Until recently this type of mange was thought to be extinct but it does occur occasionally.

Cause
A mite called *Chorioptes*.

Symptoms
There is some loss of hair, itchiness and the formation of crusts on the scrotum *(photo 2)*, the lower part of the legs particularly near the interdigital space, around the eye and on the brisket.

Treatment
Chorioptic mange is not notifiable and is not an important disease. Two benzene hexachloride baths at 14-day intervals, or four at 7-day intervals, will clear the condition.

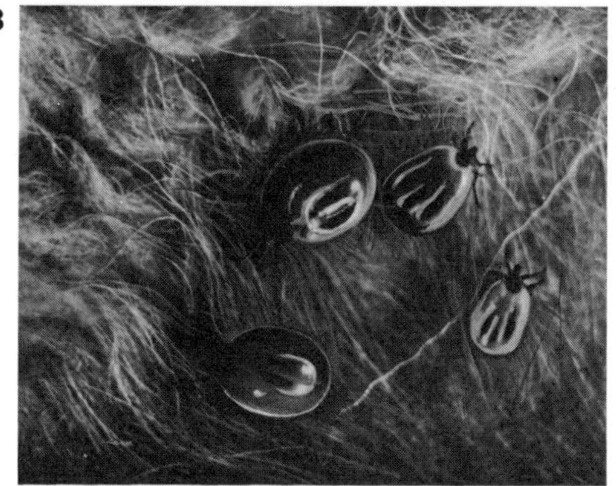

4

(d) DEMODECTIC MANGE

Although it has been recorded, demodectic mange is of no known importance in sheep.

(e) TICKS

Ticks are common external parasites of sheep in certain areas *(photo 3)*. They do not cause skin disease by themselves but are important agents for transmitting louping ill and tick-borne fever.

Regular routine dipping plays a vital part in controlling the ticks and reducing the danger of disease transmission.

(f) BLOWFLY MYIASIS *(STRIKE)*

In Britain strike usually starts when the lambs start scouring either from worm infestation or too much grass. The season for blowfly myiasis is, therefore, from the beginning of June till the end of September.

Cause
The larvae or 'maggots' of the blowflies. (Parasitic family—*Calliphorinae.*)

Symptoms
Affected lambs lose condition rapidly and are obviously restless. Close examination reveals the blowfly maggots *(photo 4)*.

Treatment
Clean off all the muddy wool and as many maggots as are visible, then immerse the lamb in sheep dip.

Prevention
Three dippings in June, July and September. The cause of the lambs scouring should be investigated and corrected.

49
Poisoning

IN MY experience poisoning is comparatively uncommon in sheep and lambs. Nonetheless, the following table of the most likely poisons *(photo 1)* with the known antidotes should be useful to all sheep farmers and veterinary students.

The most deadly plant poison encountered in sheep is the yew *(photo 2)*. There is no antidote to yew-tree poisoning.

Contrary to the general belief, lupin and charlock are non-toxic to sheep.

Obviously, it is wise to be aware of these poisons and to take considerable care that the flock has no access to them.

Poison	*Likely Source*	*Antidote*
Arsenic	Weed-killers	Freshly prepared iron oxide (ferric oxide)
Benzene hexachloride (Gammexane)	Sheep dip or insecticide used on plants	Sedatives—especially barbiturates
Carbon tetrachloride	Certain fluke drenches	Calcium borogluconate
Copper salts	Excess minerals or certain insecticides	Yellow prussiate of potassium
Lead arsenate	Insecticides	Sodium thiosulphate—intravenously
Lead	Paint—old felt impregnated with lead paint	Magnesium sulphate

1

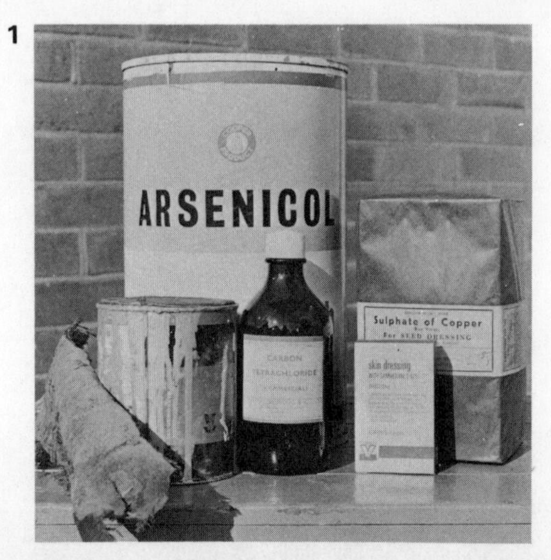

2

50
Kale Poisoning

1

RANDOM FEEDING of kale *(photo 1)* to sheep is dangerous. Given the opportunity, many sheep will eat to excess and poison themselves.

Symptoms
There is a rapid loss in condition and an acute haemoglobinuria, that is, the urine contains excess haemoglobin and is a rich red in colour *(photo 2)*.

An acute anaemia develops. The sclera or white of the eye and the inside of the mouth becomes yellowish-brown in colour *(photo 3)*, as also does the blood, which is thin and watery. If untreated, severe jaundice develops.

Treatment
Stop feeding kale immediately. Provide *ad lib* best-quality hay and inject the individual sheep with iron and vitamin B_{12} *(photo 4)*. Recovery can be spectacular, but occasional deaths are the rule.

Prevention
If feeding kale, ration it carefully to a *maximum* of 1 lb per head per day.

2

3

4

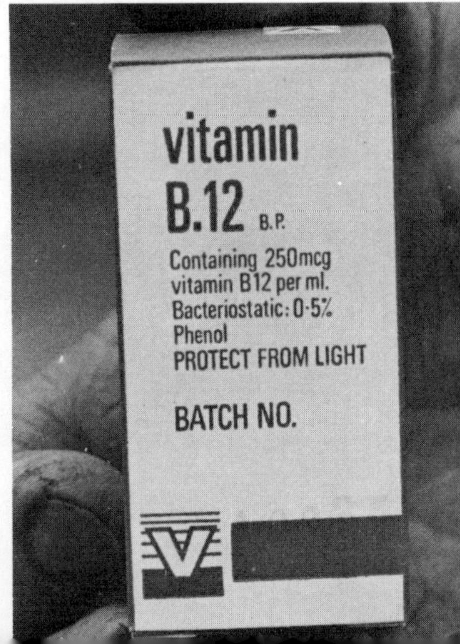

vitamin
B.12 B.P.
Containing 250mcg
vitamin B12 per ml.
Bacteriostatic: 0·5%
Phenol
PROTECT FROM LIGHT
BATCH NO.

51
Lead Poisoning

ALTHOUGH COMPARATIVELY uncommon, lead poisoning should be suspected in all cases of widespread illness or deaths, and especially where blindness or nervous symptoms are present *(photo 1)*.

Cause

In sheep the only source that I have come across is insecticide containing lead arsenate. Other possible sources are paint or old pieces of felt which are often impregnated with lead paint.

1

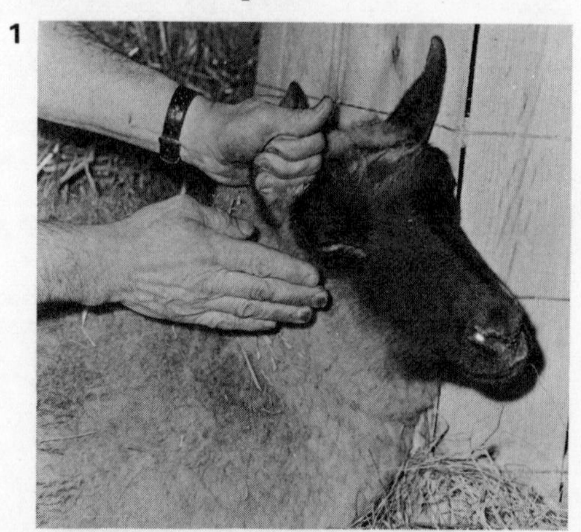

2

Symptoms

Virtually identical to those described for CCN (see p. 24). The affected sheep often stands with its head in a corner *(photo 2)*.

Treatment

Intravenous injections of sodium thiosulphate (definitely a job for the veterinary surgeon) plus drenching with Epsom salts (magnesium sulphate) *(photo 3)*—two teaspoonfuls twice daily dissolved in water and with perhaps a tablespoonful of added glucose. The magnesium sulphate is particularly useful against the lead salts (lead oxide mainly) found in paint.

Prevention

Avoid the use of insecticides anywhere near a sheep pasture, and make sure the sheep have no access to lead paint or the felt from old poultry pens.

3

52

Sturdy - Gid

(Coenuriasis)

THIS IS a brain condition, widespread throughout the country, which may affect any sheep at any time.

Cause

It is caused by the cystic stage of the tapeworm *Multiceps multiceps*, which is carried by dogs. The cystic stage is called the *Coenurus cerebralis*.

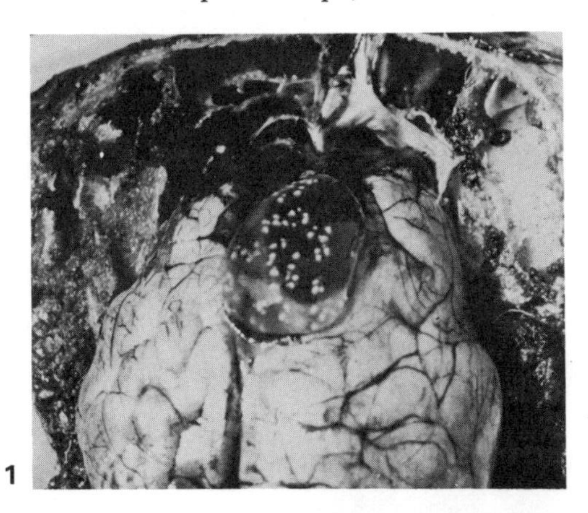

1

The sheep eats herbage contaminated with the tapeworm eggs. Inside the sheep's intestines the eggs burst to produce larvae which burrow through the intestinal wall and enter the bloodstream. Via the blood, the larvae travel to the brain or spinal canal, where they grow into cysts varying in size from a hazelnut to a small orange *(photo 1)*.

The cysts remain there until the sheep are slaughtered. Subsequently, if a dog eats the infected skull, portions of the cysts develop into adult tapeworms in the dog's intestines, and the cycle starts all over again.

Symptoms

The typical symptoms of gid appear when the cyst is developing in the brain and

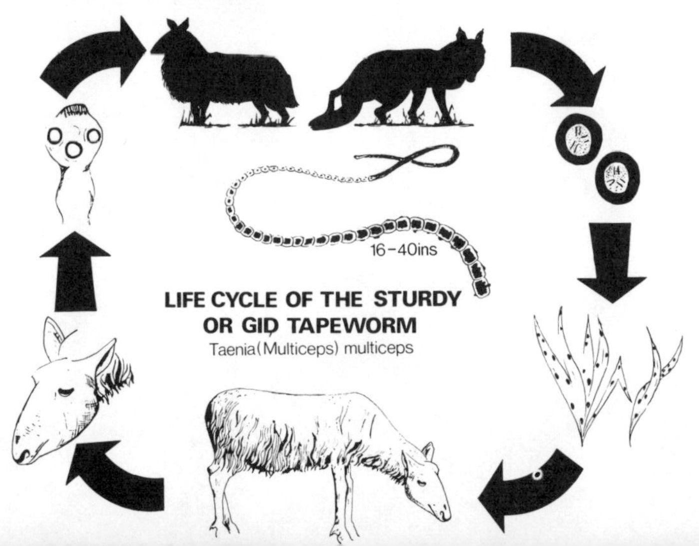

they vary according to the site and size of the cyst.

The usual first sign is holding the head to one side *(photo 2)* and circling in that direction. The eye on the opposite side may be blind.

Occasionally, the patient develops a high-stepping gait with the forelegs or a jerky walk.

When the cyst develops in the spinal canal, pressure on the cord produces progressive paralysis of one or both hind-legs.

The patient goes off food and water and separates from the rest of the flock.

2

3

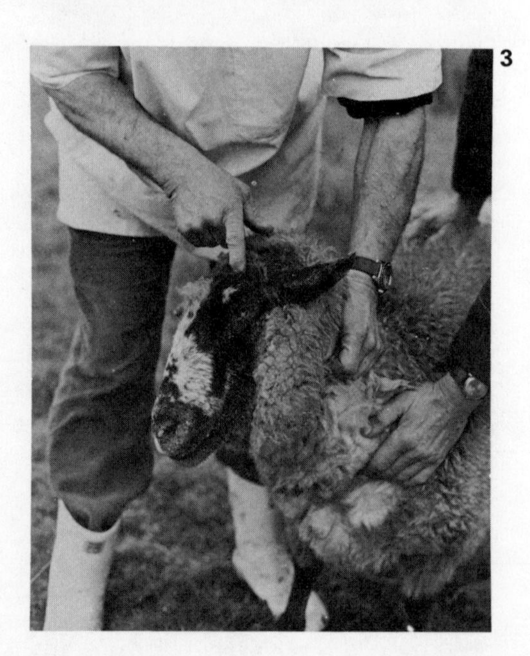

Treatment

This is a job for your veterinary surgeon. He may be able to locate the cyst *(photo 3)* and either drain it or remove it surgically.

Prevention

Dose all dogs on the farm against tapeworms at least twice a year, using the drug arecoline arsenate (your veterinary surgeon will supply this).

During, and for 2 days after, dosing confine the dogs to the kennels and burn or bury their excreta. As you can see from the diagram (p. 147), the danger arises only when the tapeworm eggs are voided on the pastures. There they develop into infective larvae which contaminate the herbage.

53
Bracken Poisoning

BRACKEN *(see sketch)* causes Bright Blindness in sheep. It does so by producing a progressive degeneration of the retina. So far the responsible factor or toxin in the bracken has not been identified.

Symptoms
The first signs appear in September, October or November, *i.e.* late in the grazing season when the pasture is sparse.

The affected sheep are permanently blind. They hold their head questioningly forward and upwards as though constantly alert *(photo 1)*.

The pupils of both eyes are circular and show little or no response to light.

Usually there is a history of accessibility to bracken.

1

There is no discharge from the eyes as in contagious ophthalmia and unlike lead poisoning, gid, pregnancy toxaemia and cerebrocortical necrosis, the vacant blindness is the only apparent symptom to develop.

Treatment
There is none.

Prevention
Burn the bracken.

54
Johne's Disease

1

2

THIS DISEASE, sometimes called paratuberculosis, is a chronic wasting disease of adult sheep occurring throughout the whole of Britain *(photo 1)*.

Cause
A germ called the *Mycobacterium johnei* or the *Mycobacterium paratuberculosis.*

How It Is Spread
The bug gets into the digestive tract in food or water contaminated by dung from affected sheep.

The *Mycobacterium paratuberculosis* produces a thickening and corrugation of the wall of the small intestine *(photo 2)*. This takes a long time; consequently a sheep is usually at least 2 years old before it begins to show symptoms of the disease.

Symptoms
The usual picture is that one or two of the flock fail to thrive. They gradually become progressively emaciated, and develop anaemia and a swelling under the jaw. The dung loses its typical pellet form but diarrhoea is rare *(photo 3)*.

Since similar symptoms can be produced by fluke and worm infestation or by pine, the differential diagnosis should be made by a veterinary surgeon.

Treatment
The disease is incurable, so all suspect cases should be slaughtered immediately to prevent further spread of infection.

Post-mortem Picture
The lining of the small intestine particularly near its junction with the large bowel (*i.e.* near the ileocaecal valve) is thickened, corrugated and golden yellow in colour *(photo 4)*. This is because the Johne's germs contain a yellow pigment.

Prevention
Get your veterinary surgeon to all suspect cases and slaughter as soon as the disease is diagnosed.

Raise the plane of nutrition of the rest of the flock *(photo 5)*. Like most other diseases, Johne's thrives in the undernourished.

Any potential reservoir of infection such as stagnant water or ponds should be fenced off, provided a fresh water supply can be provided.

3

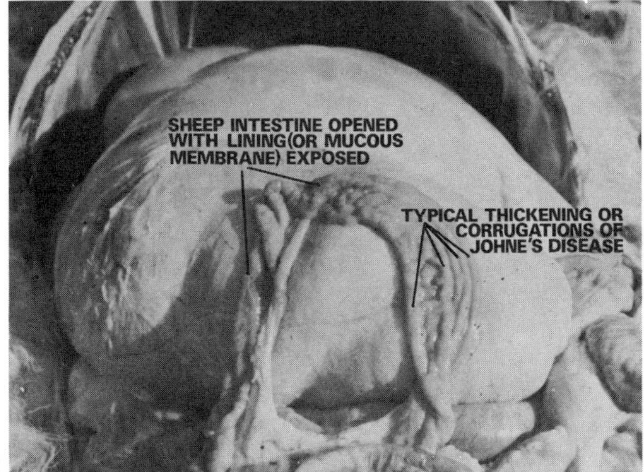

4

5

55
Lightning Stroke

IN ALL cases of sudden death, the possibility of lightning stroke should not be overlooked. Most farmers are insured against lightning, and a prompt diagnosis by your veterinary surgeon could ensure adequate compensation.

Symptoms

Sudden death. The carcase may lie alongside a wire fence or underneath a tree *(photo 1)*, but not necessarily so. In fact by far the greater number of dead animals killed by lightning stroke are found in the open field.

Post-morten Findings

A positive diagnosis can only be made by a post-mortem examination carried out by a veterinary surgeon.

The typical signs are first of all singe marks on the surface of the skin, on the shoulder *(photo 2)* and back or on the inside of the legs. The corresponding areas of the hide under the surface are clearly marked by extensive bruising and haemorrhages.

The marks and haemorrhages continue into the tissues under the skin and deep into the muscles, with the acute damage occasionally evident right through the chest and abdominal cavity.

In the chest there is invariably acute congestion and minute haemorrhages throughout both lungs, and often long strings of clotted blood in the windpipe and bronchi. The heart is contracted, with the main pumping cavities—the ventricles—nearly always empty.

The blood is not 'fluid' but it does not clot properly, the clots being soft and not clearly formed even when left for a day or two.

Sometimes a wad of grass or partially chewed hay is in the mouth, but the rest of the digestive system is absolutely normal with no signs of bloat. In fact I've never seen a bloated rumen in a true case of lightning stroke.

Obviously, it is wise to have every sudden death in the flock investigated by a veterinary surgeon.

1

2

56
Listeriosis

1

THIS CONDITION, known as 'circling disease', is uncommon in Britain but widespread in the USA *(photo 1)*.

Cause
It is caused by an organism known as the *Listeria monocytogenes*. This organism is thought to be transmissible to man, since cases of meningitis have been reported in humans engaged in coping with outbreaks of listeriosis.

Symptoms
There are two types of the disease:

The Nervous Type
The affected sheep runs a fairly high temperature and goes off its feed. It becomes weak on its legs and starts to walk in circles pushing against obstacles as though unable to see. The legs gradually become paralysed *(photo 2)*.

The Abortion Type
Abortion occurs without apparent general symptoms at any time after 3 months of pregnancy. Up to 15 or 16 per cent of lambs can be lost, and the condition can only be diagnosed by laboratory examination of dead lambs' blood and stomach contents.

In the abortion form no nervous signs appear.

Treatment
American reports suggest that sulphonamides, penicillin, streptomycin and terramycin can produce apparent cures.

Prevention
There is none.

2

57
Mycotic Dermatitis

APART FROM strawberry foot rot already described, there is one other common fungal condition which affects the skin and wool of sheep of all ages, that is 'lumpy wool disease' or 'wool rot' *(photo 1)*. It occurs more or less everywhere but is probably most common in the heavy rainfall areas.

Cause
A fungus called the *Dermatophilus congalemsis*.

Symptoms
Since there are no general disease symptoms and practically no itching or baring of the fleece, lumpy wool is not usually diagnosed until shearing time. When it is present, the wool is downgraded.

What usually happens is that patches of skin become red and moist. The acute redness subsides, and when the discharge dries, black crusts are formed which are carried up with the wool producing lumpy areas *(photo 2)* hence the name 'lumpy wool'.

I have seen the condition in lambs where it seemed to attack the regions round the ears, face and neck.

Treatment
There is no useful treatment, since the lesions are usually well established before

1

2

the condition is seen. For the moment it is probably sufficient to be able to recognise and understand the disease.

If it were possible to spot lumpy wool early it might well respond to the new oral drug *(photo 3)* now being used to treat ringworm in horses, cattle, dogs, cats and humans.

Prevention
Full-scale feeding of the new antifungal drug in flocks where the disease is known to exist would be uneconomical, though I can see a possible place for such therapy in the future, particularly if and when the cost of the drug drops to a reasonable level.

3

58
Orf

1

TWO OTHER names for orf are contagious pustular dermatitis and malignant aptha *(photo 1)*. It is a skin disease occurring in sheep and goats, but it can affect humans.

Cause

It is caused by a filtrable virus—a very minute germ which passes through a bacteria-retaining filter. The virus lies dormant in and is carried by many apparently healthy sheep; it lives in the tonsils and in other lympthatic glands.

Where And When It Occurs

It appears in all parts of the British Isles. In some areas, for example, the border counties of England and Scotland, the orf is persistent and a constant problem; in other parts the disease occurs sporadically, breaking out for no apparent reason.

Outbreaks can flare up at any time during the year but are most commonly seen in the spring and summer. Sheep of any age may be attacked, but for the most part orf is more prevalent in lambs up to 1 year old.

During an attack the virus gets into the bloodstream and travels to underneath the skin of various parts of the body. There it multiplies to form blisters, and these burst to form ulcers *(photo 2)* which become infected by secondary germs to produce pustules.

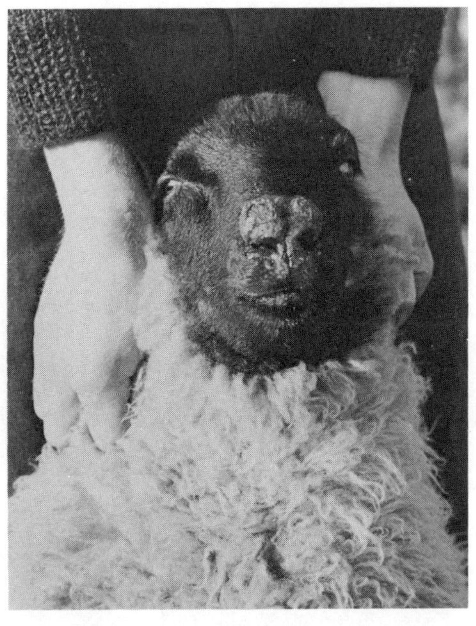

2

Symptoms

There are two types of orf—the benign type and the malignant type.

The benign type, the commonest, is characterised by most of the ulcers and pustules forming round the mouth and nostrils, especially near the commissures or junction of the lips *(photo 3)*. Occasionally, similar outbreaks occur around the coronet or tops of the feet and on the teats and udder, with the odd ulcer and pustule appearing elsewhere on the body.

Apart from being evil-looking and smelly, benign orf lesions can cause a loss in condition probably because the affected sheep have difficulty in eating. Occasionally secondary sepsis from the pustules may produce a fever and complete loss of appetite. But for the most part, the benign lesions are localised and more unsightly than dangerous.

The malignant type is a different proposition. Almost invariably it attacks sucking lambs. Here the virus invades the entire cheek cavity, producing blisters and ulcers on the tongue, cheeks, palate, gums, inside of the lips, and sometimes throughout the whole body, with similar lesions appearing simultaneously around the top of the feet, the outsides of the mouth and lips, the genitalia and the skin on the body surface.

During such a malignant outbreak in the sucking lambs, severe blisters and ulcers often develop on the teats of the ewes and these become infected, producing an acute mastitis which further complicates the whole miserable picture.

The cause of death in the sucking lambs is usually the infection by secondary bacteria, which is accelerated by the lowered resistance produced by the lamb's inability to suck or eat.

Treatment

Since there is always a possibility that the virus may become strong enough to produce the malignant type, all cases of orf—no matter how mild—must be controlled and certainly deserve careful routine attention.

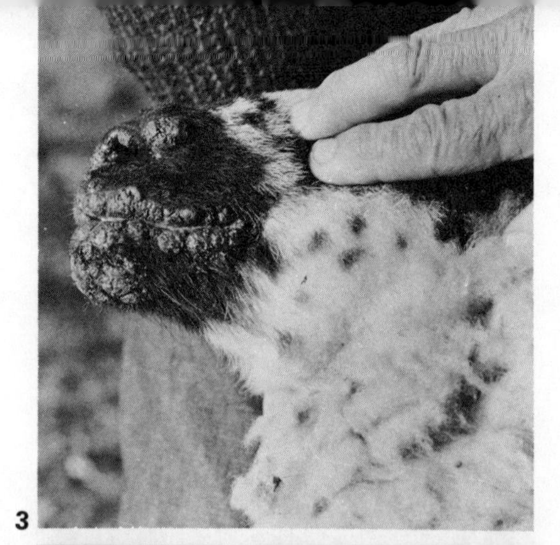

3

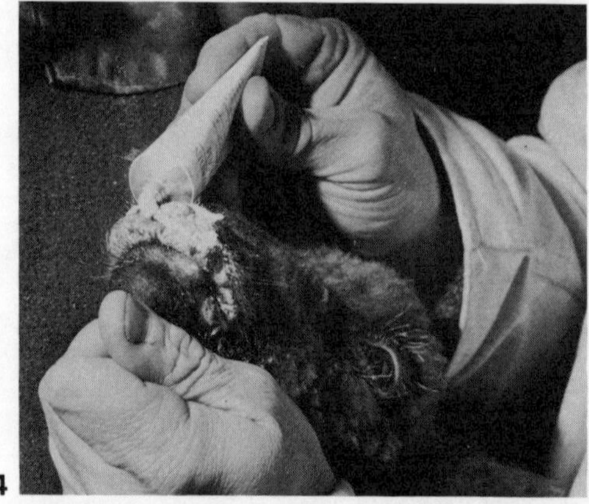

4

It is best to isolate the affected sheep immediately and to dress the lesions daily with an antiseptic, sulpha or antibiotic cream *(photo 4)* (your veterinary surgeon will be only too happy to supply the requisite preparation).

Aerosols are particularly handy for application, and I have found the chloramphenicol and gentian violet a most effective application *(photo 5)*. Oxytetra-

5

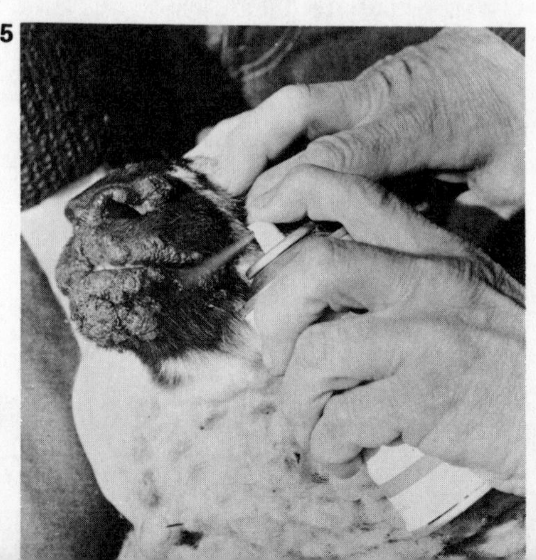

6

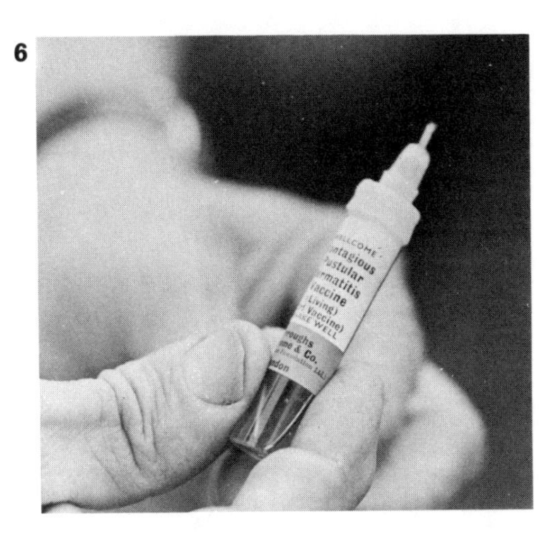

7

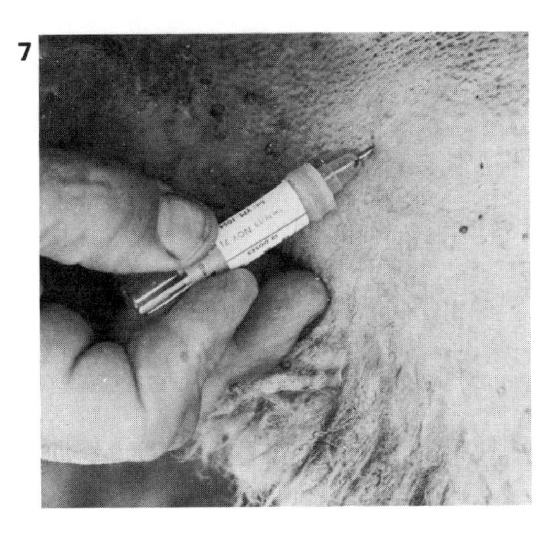

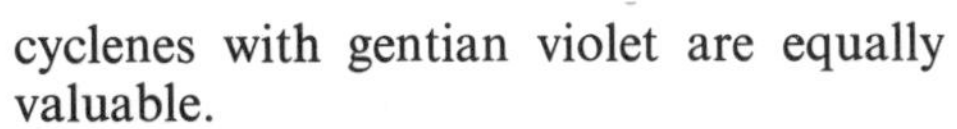

cyclenes with gentian violet are equally valuable.

Malignant orf can only be treated under the direct supervision of your veterinary surgeon.

Prevention

Fortunately, an extremely useful and inexpensive vaccine *(photo 6)* is available against orf, and in areas where the malignant type has been encountered this vaccine should be used as a routine. Elsewhere, I would say vaccination should only be resorted to after consultation with your veterinary surgeon.

The vaccine should be used at least 6 weeks before the outbreak is expected to occur (*i.e.* in March or April). This gives the antibodies against the virus plenty of time to establish themselves in the bloodstream. The vaccine must be used in sheep over 3 months of age, and vaccinated and non-vaccinated animals cannot be mixed for about 3 weeks. Ewes with young lambs at foot should not be vaccinated.

The immunity given by the vaccine lasts for 6 to 12 months, and although it is not a complete immunity, it does prevent the development of the severe widespread lesions. In other words, it does control malignant orf and reduces the lesions markedly in the benign form.

The technique of vaccination is extremely simple. Two short superficial scratches each about 1 inch long and made in the form of a cross with the vaccine-impregnated scarifier supplied with the vaccine. The site? A cleansed area of the hairless region inside the thigh *(photo 7)*. As with all vaccines, the instructions supplied should always be followed carefully.

59
Pneumonia

THE TERM 'pneumonia' simply means lack of air, and the lack of air is caused by inflammation of the lungs *(photo 1)*. Such lung inflammation can affect sheep at any age.

General Types Of Pneumonia
Pneumonia can be (1) bacterial, (2) parasitic, or (3) mechanical.

(a) BACTERIAL PNEUMONIA *(ENZOOTIC PNEUMONIA)*

This is by far the most serious sheep lung condition.

Cause
The main germ implicated is the *Pasteurella haemolytica.*

Predisposing Causes
The stress of driving to market or travelling in lorries very often flares it up.

Keeping sheep indoors in badly ventilated buildings can also trigger it off *(photo 2)*.

1

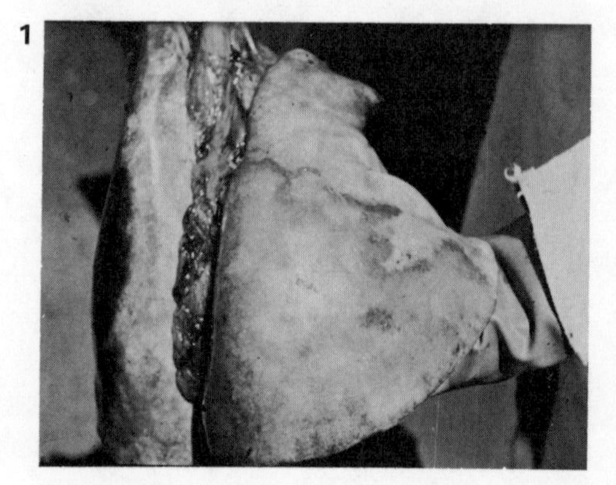

2

Sudden changes in environment particularly in the autumn may produce a severe outbreak, especially in lambs.

Symptoms
Enzootic pneumonia is an acute disease and the only symptom may be sudden death. Usually, however, the affected sheep run a high temperature (106° or 107° F): they breathe rapidly and have a nasty dry hard cough. Using a stethoscope the veterinary surgeon can detect areas of lung consolidation *(photo 3)*. They may froth at the mouth and nostrils.

Treatment
Broad-spectrum antibiotics such as oxytetracyclene given early have some bene-

3

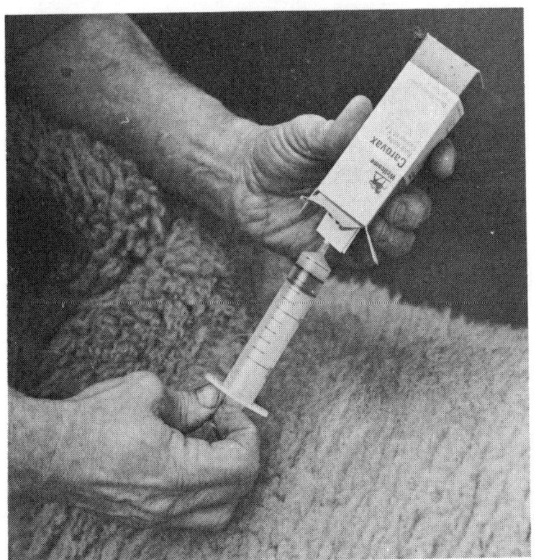

4

ficial effect in control, but the most important step is to remove any predisposing cause if at all possible. For example, if sheep are housed, improve the ventilation by providing high ridge outlets for the foul air and removing all ground draughts, at the same time, reducing the stocking rates.

Sometimes, a move to a poorer and more open grazing will apparently stop the outbreak.

Prevention
Specific vaccines are available, *e.g.* CAROVAX *(photo 4)*, but their true value has never been fully established.

If at all possible, avoid all stress factors such as markets, travelling, housing, and sudden changes in environment, particularly in the autumn when the disease is most likely to break out.

Other germs involved in sheep pneumonias are the *Corynebacterium pyogenes*, the *Actinobacillus lignieresi* and the *Staphylococcus aureus*; while chronic pneumonias can be due to lung abscesses secondary to other general infections such as tick pyaemia in lambs. But in all suspect cases it is advisable to leave it to your veterinary surgeon to establish the definite diagnosis and to prescribe the correct course of action.

(b) PARASITIC PNEUMONIA *(HUSK)*

Cause
Two species of lung worms—the *Dictyocaulus filaria* and the *Muellerius capillaris*.

Symptoms
Heavy infestations cause bronchitis. This gives rise to severe coughing which can allow parasitic or bacterial pneumonia to develop. There is usually a discharge from the eyes and nose.

Treatment
Consult your veterinary surgeon at once. He will diagnose the condition by examining samples of dung and will prescribe the specific treatment *(photo 5)*.

5

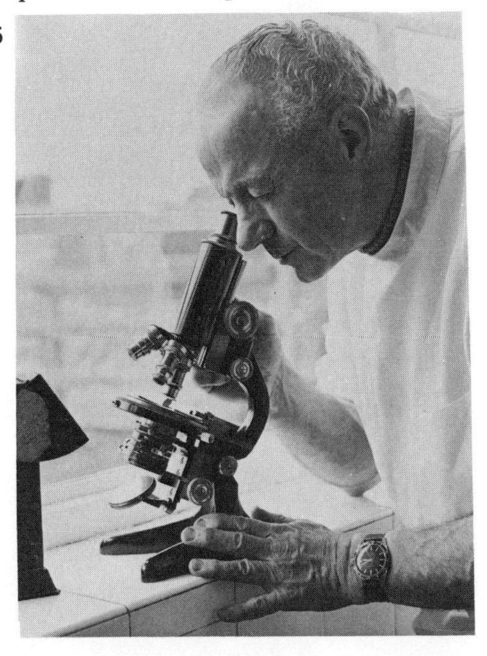

Fortunately husk in sheep rarely becomes the serious problem that it does in calves, but nonetheless any outbreak of coughing should be investigated by your veterinary surgeon.

(c) MECHANICAL PNEUMONIA

All too frequently pneumonia is due to the misuse of the drenching gun *(photo 6)*. *In careless or ignorant hands the dosing gun is a lethal weapon.* What happens is that the liquid worm or fluke remedies are pushed into the windpipe causing choking, drowning or, if the victim survives long enough, pneumonia.

Treatment
Casualty slaughter for food is undoubtedly the best treatment.

Prevention
If a drenching gun has to be used, the sheep's head should be kept horizontal and the nozzle gently introduced—not into the back of the throat, but merely over the bulge at the rear of the tongue *(photo 7)*.

Fortunately, most anthelmintics can now be given hypodermically and, in my opinion, this is by far the easiest and best method. Certainly it will completely prevent the many mechanical pneumonias resulting from drenching.

There are two other pneumonias of sheep, both chronic pneumonias and both seen mainly in sheep in Iceland. They are called *jaagsiekte* and *maedi*.

Jaagsiekte has been reported in Britain among older sheep. It is a tumour-type condition believed to be caused by a virus. Since there is no treatment and no vaccine, the only logical control is to slaughter affected animals as soon as they are detected.

6

7

Obviously, with all lung conditions in sheep, the best policy is to consult your veterinary surgeon. He, probably with the help of a scientist, will establish the correct diagnosis and will draft out the best procedure to follow.

60
Scrapie

SCRAPIE IS a chronic nervous disease of adult sheep *(photo 1)*.

Cause

The cause is unknown but it may be a filtrable virus which settles in the central nervous system (the brain and spinal cord) and in the spleen.

The disease appears to be spread by the pasture or by congenital infection through either the ewe or the tup, though both may appear healthy and normal at the time.

It attacks when the ewes or tups are anything from $1\frac{1}{2}$ to 9 years old, and it takes from 18 months to 3 years for the symptoms to appear.

The disease occurs in Scotland, certain parts of England, and in Germany and France.

It would appear that certain breeds are more susceptible than others because scrapie is most commonly seen in the Border Leicester, the Suffolk, the Cheviot, and in the half-bred crosses from these breeds.

1

2

It is rare in the Scottish Blackface and in the Blackface crosses.

Some years up to 20 per cent of the flock may be affected; other years there may only be an occasional case.

Symptoms

The first sign is usually itching called pruritis. The itching may be mild, but usually the affected sheep rub themselves repeatedly against fences, troughs, posts, etc. *(photo 2)*. As the disease progresses, the persistent itch causes the lips to twitch and the tail to wag.

The suspect sheep becomes hyperexcitable. Shivering of the head and neck may cause slight nodding, and when the animal is chased or rounded up, these nervous symptoms become more marked; the sheep in fact may have difficulty in walking and may fall down in a fit *(photo 3)*.

The itching interferes with eating and sleeping and leads to a marked loss in condition. The wool becomes lighter in colour and eventually emaciation and paralysis supervene. Death occurs in anything from 2 weeks to 6 months.

Treatment
There is none.

Prevention
So far none, but considerable research work is going on and there is a chance that a vaccine may eventually be produced.

3

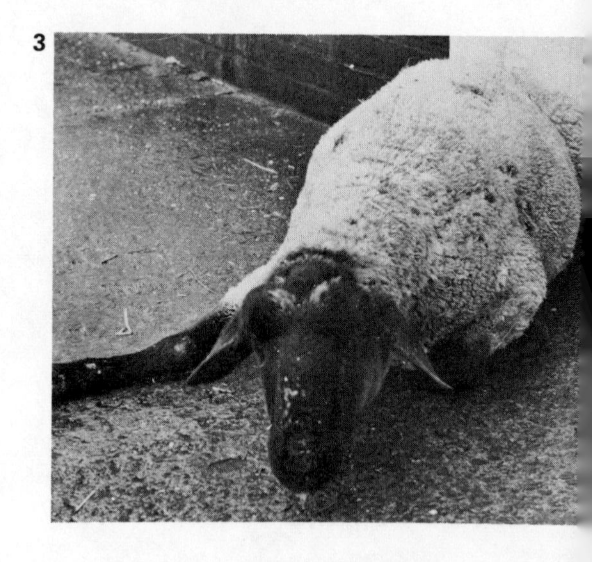

61
Sheep Nasal Fly Disease

THIS IS a summer disease which apparently affects sheep in many parts of England but which, so far, has not been recorded in Scotland *(photo 1)*.

Cause
It is caused by the larvae of a fly called the *Oestrus ovis (see diagram)*.

The adult fly lays its larvae in the nostrils of the sheep.

The larvae feed on the nasal mucous membrane for up to 9 months, then migrate up into the sinuses of the head, undergoing two changes and eventually passing down again into the nostrils to be sneezed out on to the pastures, where they develop into adult flies.

Symptoms
The majority of infestations are by only a few larvae and show little or no sign.

With heavy infestations the affected animal shakes its head and rubs its nose on the ground.

The nostrils develop a pusy discharge *(photo 2)*, and the patient grazes less and loses weight.

OESTRUS OVIS
MATURE THIRD LARVA

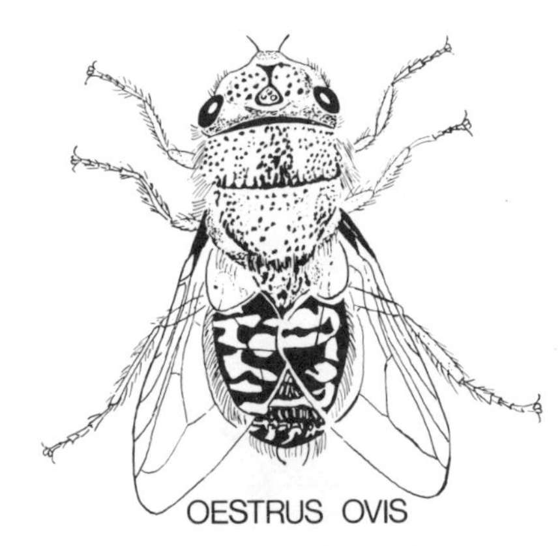
OESTRUS OVIS

Treatment
There is none. Steaming of the individual with medicated inhalations will help recovery.

Prevention
None known.

1
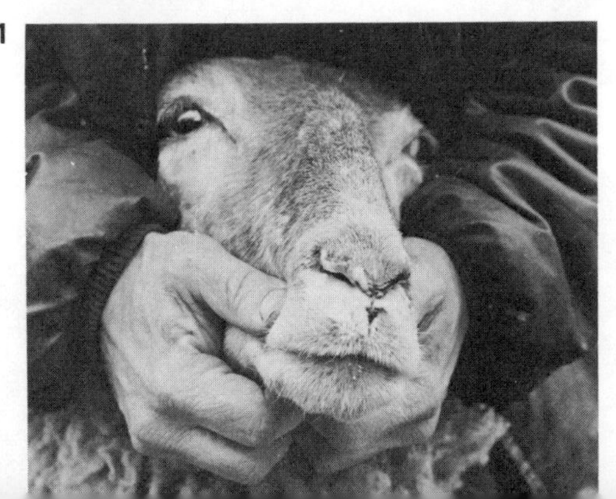

2
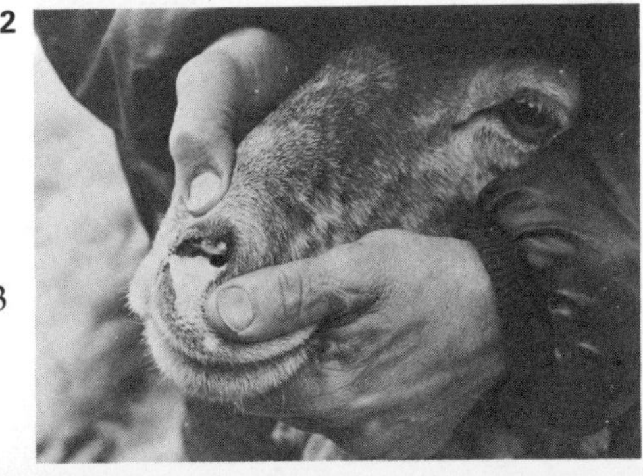

62

Protecting the Indoor Ewe and Lamb

LIVING OUTDOORS has a great deal to commend it, and for countless years sheep, by doing so, have avoided the disease problems of intensification. In recent years, however, there has been a move towards housing the ewes and lambs *(photo 1)* and naturally this has produced special complications. It is vital, therefore, to be aware of the dangers and to take every possible step to control and prevent them.

1

WORMS present no problem in lambs housed from birth (this is obviously one great advantage of the system), but they can cause deaths in older ewes, especially when they are not being very well fed.

The routine prevention in all housed enterprises is to dose *all* the ewes when they are first housed, and to get your veterinary surgeon to check dung samples a month after dosing.

COCCIDIOSIS may start in the housed lamb about a month after birth, causing a rapid loss in condition.

Routine prevention involves dung checks on the ewes, and possibly preventive medication of the feed and drinking water.

LICE are the outstanding cause of itching among housed ewes, especially if the ewes are on a low plane of nutrition.

To avoid this trouble, dip all ewes before housing, and if the food is sparse, dip them again later on, in January or in February.

PREGNANCY TOXAEMIA is prevented in the same way as for outdoor sheep, that is, by a steady increase in feeding and condition during the 8 weeks before lambing; but control is made easier in housed ewes by grouping and penning the ewes according to their lambing date. If the lambing dates are not known, then group them by weight. All ewes that appear to be losing condition should be fed and housed separately.

2

PNEUMONIA AND BRONCHITIS. Provided the diet is adequate and consistent, the most important successful preventative of pneumonia and bronchitis is correct ventilation in the house *(photo 2)*; so before you consider housing your ewes, get the house 'vetted' in the true sense of the word. Have no fear; your veterinary surgeon will not hesitate to consult a specialist if he has any doubts.

COPPER POISONING. This is a possible hazard when concentrates are fed to housed sheep. Keep veterinary supervision of the copper content of your ration and *never, under any circumstances, feed minerals containing copper.*

ACIDOSIS. This is a condition where the housed ewe goes off her food almost completely. It is seen chiefly when barley is first fed as a concentrate and insufficient fibre is available. Obviously, the simple answer here is to get the ewes inside and used to eating hay before starting the 8-week build-up of concentrates containing as little barley as possible or none at all. It has been my experience that barley is a menace to the pregnant and lactating ewe, consistently producing metabolic and digestive disturbances.

INFLAMMATION OF THE RUMEN OR FIRST STOMACH. This is seen in the growing lamb fed on barley concentrates with little or no roughage. It causes colicky pains and death. The answer? Don't feed barley and, *as a routine, always provide* ad lib *good quality hay*. With *ad lib* hay you'll probably get away with feeding a percentage of barley, but you are much better without it.

INFECTIOUS DISEASES GENERALLY. When planning the house, aim to have the ewes penned in small numbers. The ideal is around 12 ewes per pen, with 25 the maximum.

Then vaccinate the ewes against all the enterotoxaemic diseases and against tetanus, blackleg and black disease, with a booster dose of the vaccine or vaccines immediately before lambing.

ORF. Although, as stated in the 'orf' chapter, the vaccine is not 100 per cent effective, it does avoid an outbreak of the malignant type which can take three-quarters of the lamb crop. So I advise routine orf vaccination in all housed ewes. The vaccine is cheap, and your veterinary surgeon will supply it and instruct in its application.

ECZEMA AROUND THE EYES. This is due to a contagious virus, probably the herpes virus, which seems to spread when the ewes are troughing. Control it by penning in small numbers and prompt isolation and treatment.

ABORTION. This is the greatest potential hazard to the pregnant housed ewe. Apart from the traumatic miscarriage, there are two main types:

Enzootic Abortion. This is now found all over Britain. Consequently, vaccination against it is an absolute must in all housed ewes. Even vaccinated ewes may occasionally abort. If they do, they should be penned separately at subsequent lambings.

Vibrionic Abortion. This is due to a vibrio foetus organism, which is picked up by the mouth, and is not spread by the ram. So far, there is no satisfactory vaccine against this, but there is every prospect that one may soon be available. In the meantime, the penning of the ewes is the most effective control weapon.

One important point: recovered ewes have a strong immunity and should be kept—not discarded. Also, as with enzootic abortion, they should be penned separately at subsequent lambings.

SALMONELLOSIS. There are several types of the germ which can cause abortion, enteritis or death. So far, the sheep infection is apparently limited to certain areas, and the simple preventive measure in those areas is *not to house sheep under any circumstances.*

CONTAGIOUS OPHTHALMIA (*i.e.* the eye infection caused by the Rickettsia). Obviously, this is much more likely to spread when the ewes and lambs are bunched together. To control, isolate at once and give immediate treatment with antibiotic (preferably chloramphenicol) preparations.

MASTITIS. This is much more likely to occur in housed ewes, and it is more difficult to spot the early symptoms *(photo 3)*. It is seen chiefly when the lambs are weaned at about 6 weeks old. One way of preventing the mastitis is to wean gradually, allowing the lambs to suck for 10 minutes once a day for a week or two, then once every second day. As in cattle, there is no efficient vaccine.

FOOT ROT. If the ewe's feet are carefully treated before housing, this disease is not likely to become a major problem. Nonetheless I advise the use of the foot rot vaccine, as well as feet care, before the sheep are penned up.

For the new-born and young lambs the two greatest hazards of housing are navel infections and *E. coli*, though to some extent the losses from these in housed lambs are balanced up by the number saved by being born in comparative warmth.

3

NAVEL INFECTIONS. These lead to septicaemia and joint ill. The most effective control measures are, first of all, to try to lamb each pen down in as short a period as possible. This avoids the build-up of infection on the pen floors. Secondly, dress the navels of the lambs with iodine or antibiotics as soon after birth as possible.

E. COLI INFECTION. Sometimes this causes diarrhoea when the lambs are 2 or 3 days old, but usually what happens is that the lamb stops sucking about 12 to 36 hours after birth and dies a few hours later. The control? Prompt treatment by oral and parenteral antibiotics. This calls for close liaison with your veterinary surgeon, who will have your coli carefully typed. Also, as with navel infection prevention, aim at lambing the entire pen in as short a period as possible to prevent disease build-up.

Where there is a history of the disease, dose each lamb by the mouth with an appropriate antibiotic before the lamb is 12 hours old.

In all cases—as soon as evidence of navel infection or *E. coli* appears—remove the ewes still to lamb to a clean pen immediately.

It must be quite obvious that lambing ewes indoors is *not* an enterprise to be undertaken lightly. The owner, the shepherd and the veterinary surgeon must play his part, but the vet must be in on the entire project right from the start. If, as so often happens, he is consulted only after trouble appears then losses can be catastrophic.

In the housing of all sheep, particularly the breeding ewe, preventive medicine planned and prescribed by the veterinary surgeon is the most essential economic factor in the entire enterprise.

63

Routine Preventive Medicine

Before and After Lambing

IN NO other class of animal has preventive medicine *(photo 1)* been developed to such a practical and successful extent. In fact with the modern armoury of knowledge, vaccines and drugs the majority of sheep losses must be the responsibility of the supervising veterinary surgeon.

First of all, he can and should ensure against any deaths from pregnancy toxaemia or lack of milk in the ewes by ordering simple, good feeding starting at least seven weeks before lambing and building up the quantity as the pregnancy progresses. The drugs, cortisone and multivitamin *(photo 2)*, should always be at hand to treat promptly any suspect cases, bearing in mind that blindness and abnormal behaviour attributed to pregnancy toxaemia can be due to cerebrocortical necrosis brought on by a deficiency of thiamine or vitamin B_1; and that they can also be due to lead poisoning.

2

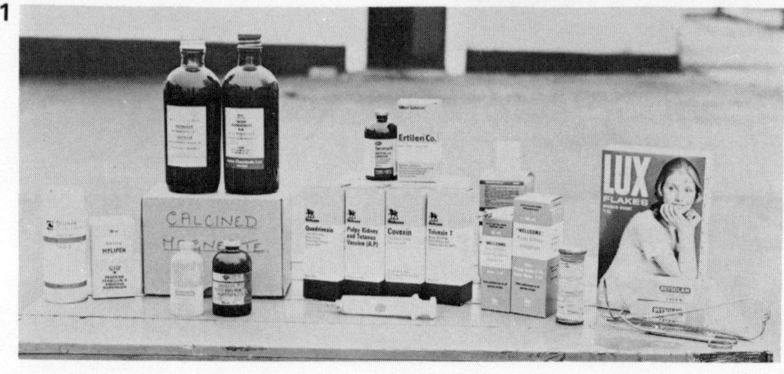

1

3

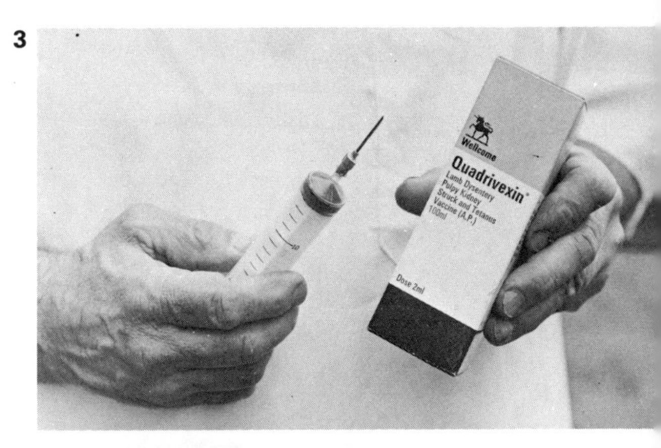

4

Hypomagnesaemia losses are avoided by the incorporation in the feed of the correct amount ($\frac{1}{4}$ oz per head per day) of calcined magnesite either directly or in magnesium nuts, or by the provision of magnesium acetate in molasses. If cases do occur, sterile magnesium injections should be available for immediate treatment.

Lambing sickness seen in ewes just before, during, or after lambing can be prevented or at least treated promptly by the injection of 80 to 100 cc of a reputable calcium solution.

All the clostridial diseases and infections—enterotoxaemia, pulpy kidney, struck, lamb dysentery, braxy, black disease, tetanus, blackleg (black quarter), malignant oedema and post-parturient gas gangrene—can and should be prevented by vaccinating the ewe flock with the appropriate multivaccine *(photo 3)*. Twice the first year—one month and again immediately before lambing, and once each subsequent year just prior to lambing. This not only keeps the adult flock clear of enterotoxaemia, braxy, black disease, tetanus, black quarter, malignant oedema and gas gangrene, but ensures that the lambs have protection against pulpy kidney, lamb dysentery and tetanus during the first vital eight to twelve weeks of life. If the lambs are going off for slaughter by that time, then there is no danger. However, all lambs kept over three months, whether as replacement breeders or fatteners, *must* be vaccinated. The ideal is to give them their first dose at six weeks and their second at twelve weeks.

For additional protection, concentrated lamb dysentery serum (which contains antibodies against pulpy kidney as well) can be given to the lambs at birth, while the pulpy kidney immunity can be temporarily boosted by a dose of antiserum at six, seven or eight weeks.

Lambing fatalities can and should be kept to a minimum by patience, thorough cleanliness and commonsense. A few moments to apply an antibiotic cream to the hand and arm for even the most casual examination, profuse lubrication with soap flakes and warm water for all obstetrical interference, and an antibiotic cover after every difficult lambing and during any retention of the placenta, should virtually eliminate obstetrical losses.

To combat prolapse problems, a good pair of needle holders, a decent needle and some nylon make stitching jobs a simple matter, but the kit should be smothered in

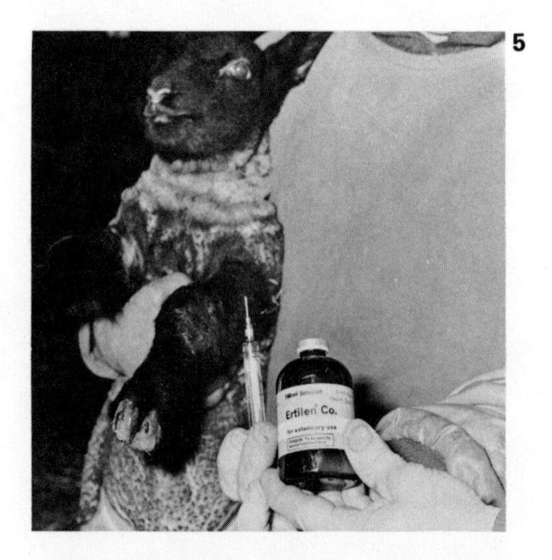

5

antibiotics during use. All stitched patients should be provided with a long-acting (5 to 7 day) antibiotic cover to prevent infection of the suture wounds.

Joint-ill can be fought off by hygiene and general cleanliness. Lambing pens should be bedded down with clean straw between each patient and the lamb's navels should be dressed as soon after birth as possible. The most convenient and perhaps the best dressing is the aerosol spray containing chloramphenicol, terramycin or aureomycin *(photo 4, page 169)*.

Another wise precaution is the dipping of the shepherd's feet in antiseptic solution before entering the lambing pens.

Diarrhoea or scour in vaccinated lambs should be treated immediately with oral sulpha drugs and antibiotics. It is most likely due to an *E. coli* infection. As in calves, there occurs a build-up of this infection in pens, so as a routine prevention the ewes should be moved on to a fresh lambing area half way through lambing, or better still, move them twice during the lambing period.

On farms where Nematodirus is known to exist, it is probably best to anticipate the trouble by preventive dosing at the beginning and end of the first danger month (*i.e.* March or April in Northern Ireland and May in England), with perhaps a third dose at the end of June. This triple dosing may be expensive but nothing like as expensive as an active infection could prove.

With the continual improvement in the modern anthelmintics, a single early spring dosing may be sufficient to prevent the majority of outbreaks.

Another useful preventive hint is not to use the same pasture for lambs in successive years.

With the ordinary gastro-intestinal helminthiasis it is obvious from the life cycle that clean sheep can only become infected by eating the infective larvae from the pasture. Maximum pasture contamination occurs in the spring due to the so-called 'spring rise' in the output of eggs. At that time an apparently healthy sheep may pass out over eight million eggs daily.

The basis of control, therefore, must be to avoid a build-up of infective larvae on the pastures.

The first job is to kill the adult worms in the sheep. Using one of the modern anthelmintic injections, routine dosing should be started in the spring, about a fortnight before the later lambing. This will reduce the 'spring egg rise' to a minimum.

The lambs should be injected at six weeks and once every six weeks until the autumn, when the entire flock should be done again.

The dosing should be combined with sensible pasture husbandry. If the sheep are folded on small areas, for example, the hurdles should be moved every week at least since it takes three to seven days for the larvae to become infective. Where ample grazing is available, the grazing should be changed after each dosing.

Plouging-in will not kill the larvae, but if another crop is taken off the land before grass is re-sown the land should be clear.

One sensible way of cleaning up pastures is by cross grazing them with cattle or horses; the sheep larvae cannot develop in these animals and are destroyed.

So far no pasture dressing has been found that will kill parasitic larvae without destroying the herbage.

In fluke areas control measures should be started during the winter months. If the summer has been warm and wet, then the entire flock should be injected once a month from October to a month or six weeks before lambing. This monthly dosing will destroy each batch of adult flukes shortly after they arrive in the bile duct, before they have had time to do a great deal of damage to the sheep's health, and more important perhaps, before they reach full sexual maturity. This will prevent further contamination of the ground since the immature adults will not produce any eggs.

Another way of tackling the problem is to reduce the snail population, and the best long-term way of doing this is by drainage of the danger areas.

Where the snail areas are comparatively small, they can be fenced off and the snails destroyed by a molluscicide *e.g.* copper sulphate, or the more effective Frescon (Shell Chemical Ltd). The best time to apply the molluscicide is in the early spring—March, before the snail breeding gets going with a second dressing in June before any infected snails can be shed their cercariae.

The ideal, of course, in bad fluke areas is to do all three—drain, dress with a molluscicide, and dose with anthelmintics.

These are but a few better known preventive routines. Many other diseases like orf, louping-ill, enzootic abortion, enzootic pneumonia, swayback, pine, double scalp, skin diseases etc. can be dealt with promptly and efficiently in the areas of their occurrence. (See appropriate chapters.)

For pneumonias and any other more complicated conditions the broad spectrum antibiotics *(photo 5)* are available, through the veterinary surgeon, to all modern flockmasters.

All in all, we veterinary surgeons are extremely fortunate to have available such potent defences against sheep diseases. It says a great deal for the application and efficiency of our research workers over a long period of time. I often think their efforts are not sufficiently applauded.

64

Preventive Medicine Plan

On Lowland Farms Free from Ticks

1

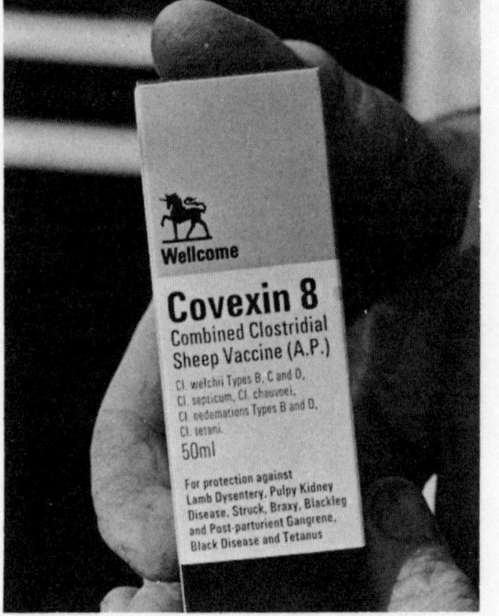

2

HERE IS a general plan for farmers and shepherds to follow:

Two Months Before Lambing
Provide 5 per cent copper licks to ewes as a prevention against swayback *(photo 1)*. Give first injection against fluke in fluke areas.

One Month Before Lambing
EWES
Inject multivaccine against all diseases *(photo 2)*, with a second dose immediately before lambing if first year on the farm. In subsequent years give a booster dose immediately before lambing.

ALL THE SHEEP ON THE FARM, INCLUDING TUPS
Dose against stomach and bowel worms and against fluke if in a fluke area (second injection).

New-born Lambs
Spray navels with chloramphenicol or oxytetracyclene to prevent navel infections.

If vaccination of ewes is in doubt inject the new-born lambs with lamb dysentery serum but this should not be necessary if a reputable polyvalent vaccine has been used on the ewes.

Three- to Four-week-old Lambs
Pulpy kidney vaccine (enterotoxaemia).

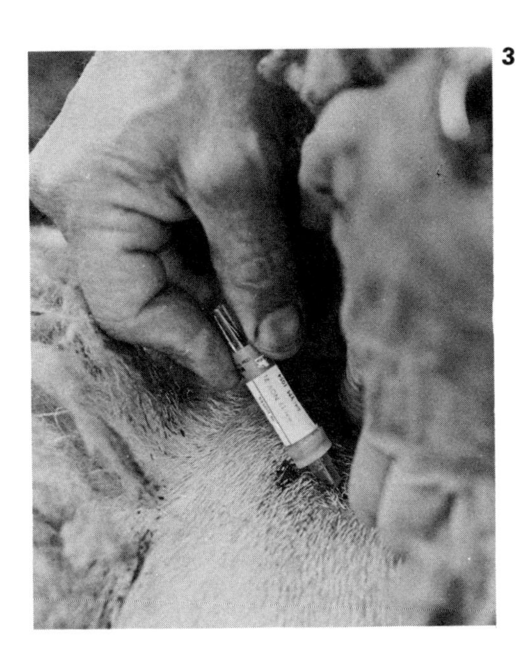

3

Three-month-old Lambs
Orf vaccine on problem farms only *(photo 3)*.

May to June
Crutching to prevent blowfly strike.
Dipping to prevent blowfly strike.

July to August
Compulsory dipping in some counties.

August to November
ALL STOCK
Compulsory dipping.

ALL BREEDING STOCK
Virus abortion vaccine—1 month before tupping (in problem areas only).

ALL STOCK
Booster dose of polyvalent vaccine.

65
Hill Sheep Farming in Scotland

THE SCOTTISH sheep industry is based on a three-tier system, and some knowledge of this system is essential in order to understand the disease control that is practised.

On the higher hills which form the top tier *(photo 1)*, Blackface and Cheviot flocks are bred pure and the only 'bought in' sheep are replacement rams. The ewes remain on the hill throughout the whole of their productive life, that is, until they are $5\frac{1}{2}$ years old or until they have produced their first four lambs. They are then sold or 'cast'.

Lambing occurs on the hill around mid-April, and the lambs remain with the ewes until they are weaned in September.

The 'cast' Blackface and Cheviot ewes are sold in the autumn and are bought by sheep farmers on the middle tier, that is, farms comprising the lower slopes of a hill farm or a better hill farm at a lower altitude. There the ewes are crossed with a Border Leicester ram; this produces a Greyface lamb with the Blackface and a Halfbred lamb with the Cheviot.

Lambing on these marginal farms begins in the middle of March.

The lowest or third tier comprises the lowland or arable farms. There the Greyface and the Halfbred ewes are crossed with Oxford Down or Suffolk rams to produce early fat lambs for slaughter. On these lowland farms lambing occurs in January or February.

On the top tier the replacement ewes come from the previous year's lamb crop. The chosen ones spend 6 months—from the end of September to the beginning of April—on a lowland farm. On 1st April they are dipped in an anti-tick dip, then turned out on the part of the hill on which they were born.

1

66

Preventive Medicine Plan for Hill Sheep

THIS DIAGRAM, prepared by Mr Alex Wilson, MRCVS, Veterinary Investigation Officer at Auchencruive, sets out clearly the preventive routine for all hill sheep farms to follow.

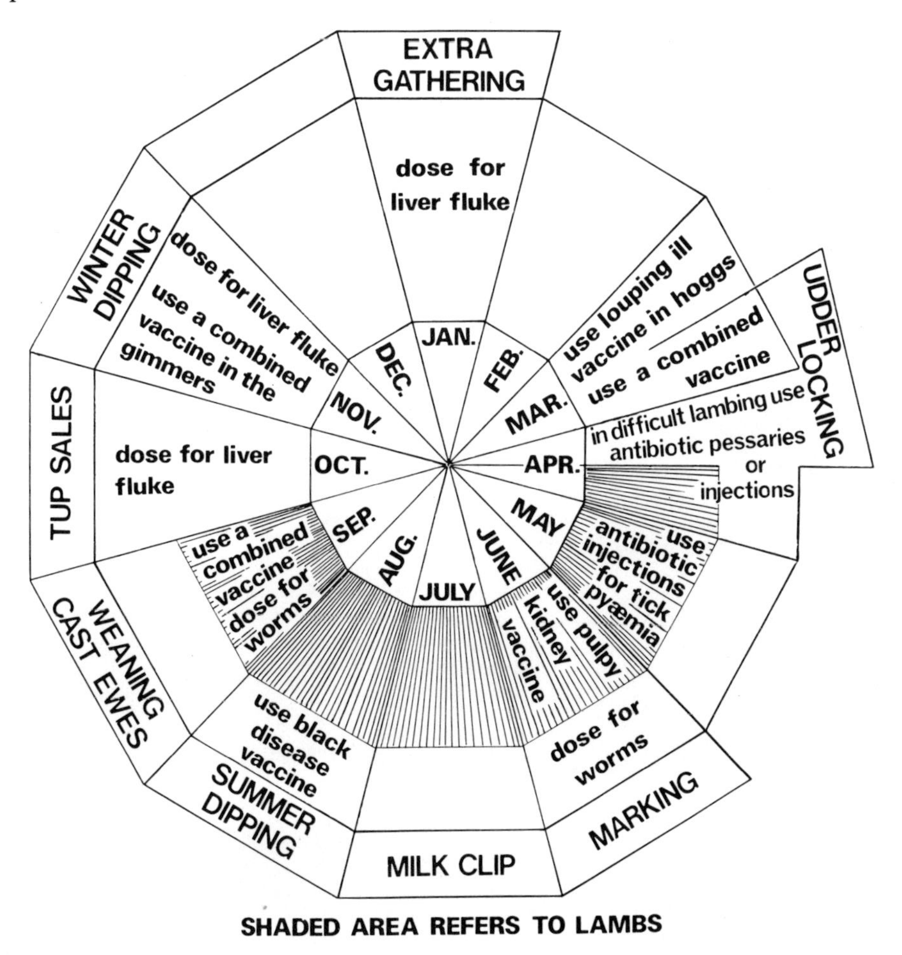

SHADED AREA REFERS TO LAMBS

Index